FUN WITH MATH

PRINCETON ■ LONDON

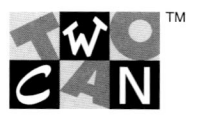

Published in the United States and Canada by
Two-Can Publishing LLC
234 Nassau Street
Princeton, NJ 08542

www.two-canpublishing.com

© 2001 Two-Can Publishing

For information on Two-Can books and multimedia,
call 1-609-921-6700, fax 1-609-921-3349, or visit our Web site at
http://www.two-canpublishing.com

All rights reserved. No part of this publication may be reproduced,
stored in a retrieval system or transmitted in any form or by any means electronic,
mechanical, photocopying, recording or otherwise, without prior
written permission of the publisher.

'Two-Can' is a trademark of Two-Can Publishing
Two-Can Publishing is a division of Zenith Entertainment plc,
43-45 Dorset Street, London W1H 4AB

ISBN 1-58728-0523

1 2 3 4 5 6 7 8 9 10 02 01 00

SHAPES, MEASURE, PATTERNS & GAMES
Consultants: Wendy and David Clemson
Editor: Diane James
Photography: Toby
Text: Claire Watts

SHAPES & MEASURE
Editorial Assistant: Jacqueline McCann
Design: Beth Aves

Printed in Hong Kong by Wing King Tong Company Limited

CONTENTS

Shapes Around You	6	Sand Timer	64
Printing Shapes	8	Tangrams	66
Making Shapes	10	Looking at Patterns	70
Shape Pictures	12	Spiral Snakes	72
Patches	14	Beads	74
Cutout Shapes	16	Cake Fun	78
Slit-and-Slot Shapes	18	Weaving	80
Stand-Up Jungle	20	Dot Patterns	84
Food Monster	24	Cut Paper	86
Looking at 3-D Shapes	26	Tiles	88
Make a Cube	28	Mosaics	90
Desk Organizer	30	Wrapping Paper	92
Build an Animal	32	Mirror Prints	94
Stretchy Jewelry	34	Fun Clothes	96
What Size Is It?	38	Stationery	98
How Tall Are You?	40	Playing Cards	102
Puppet Play	42	Number Cards	104
Finger Puppets	46	Memory Game	106
Cutout Mask	48	Bingo	108
Paper Dolls	52	Dominoes	112
Bottle Band	54	Spinners and Dice	116
A Mobile	56	Beetle Game	118
Baking Cookies	58	Board Games	122
Cookie Boxes	60	Snakes and Ladders	126
Wrapping Boxes	62	Index and Notes	128

Shapes

6 Shapes Around You

Everything you look at has a shape. Some things are curved. Some are straight. Some have points or corners. What words do you use to talk about the shapes around you?

Look at all the shapes on these pages. How many curved edges does each one have? How many straight edges does each one have?

Playing with Shapes

You can put some shapes side by side to make a completely different shape. You can also make patterns with shapes. Experiment with different shapes of building blocks.

The activities in this book will help you:
- spot and name shapes around you.
- learn some important facts about solid and flat shapes.
- make patterns using shapes.
- learn how to build and take apart shapes.

8 Printing Shapes

● Look for objects with interesting shapes that can be used to print a pattern or a picture. To print the pictures shown below, we used a sponge, a cork, an eraser, and a wooden building block. Find some other shapes to print. Get permission to use them.

● Cover one side of your shape with paint. You could use a paintbrush or dip the shape into the paint. Press the shape carefully onto the paper.

● Some shapes can fit together with no gaps in between, like these squares. Which of your shapes fit together? Which shapes do not fit together?

Here's what you learn:
● how to recognize different shapes.
● how to fit shapes together.

10 Making Shapes

Here are some simple ways to make basic shapes from colored paper.

Square
- Start with a rectangle of paper.
- Place the paper so that one of the short ends is toward you.
- Fold the top left corner toward you until the top of the paper lines up with the right-hand side of the paper.
- Cut off the single piece of paper left at the bottom.
- Unfold the paper and you will have a square.

Triangle
- Cut your square right down the diagonal fold to make two triangles.

Circle
- To make a circle, find a round object, such as a jar or can. Put it on a piece of colored paper and trace around it.
- Starting from the edge of the paper, cut out the circle carefully.

Here's what you learn:
- how to recognize different shapes.

Semicircle
- Fold your paper circle in half. Cut along the fold to make two new shapes. The new shapes have a curved edge and a straight edge.

12 Shape Pictures

We made the picture here using colored paper shapes. You will be surprised at how easy it is to make an interesting picture of your own. Cut or tear many different paper shapes before you start.

Try It Out
● Arrange your shapes on a sheet of paper to make a picture.

● Don't glue the shapes down right away. Experiment with the shapes and move them around. When you are happy with your picture, carefully lift up each shape and glue it in place.

● Circles and ovals are good shapes for making heads and bodies. Triangles, rectangles, and squares are good for making buildings.

Here's what you learn:
● how to fit shapes together.
● how to use different shapes in design.

14 Patches

With just a few old scraps of fabric, you can make a colorful patchwork cloth.

● The six-sided shape below is called a hexagon. Trace it onto other paper and cut it out. Make lots of paper hexagons.

● Place each paper hexagon on a piece of fabric. Cut around the paper, leaving extra fabric on all sides.

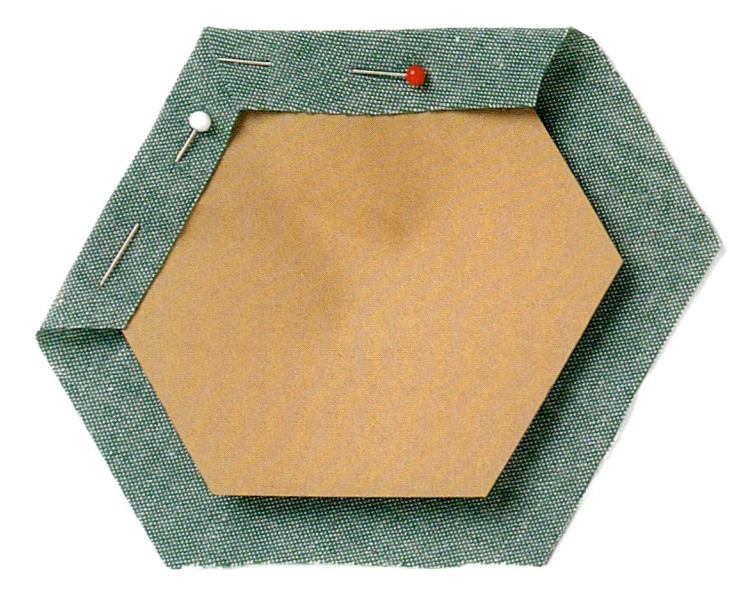

● Fold each edge of the fabric and ask an adult to help you pin it to the edge of the paper. When you have pinned several pieces, you are ready to sew the hexagons together!

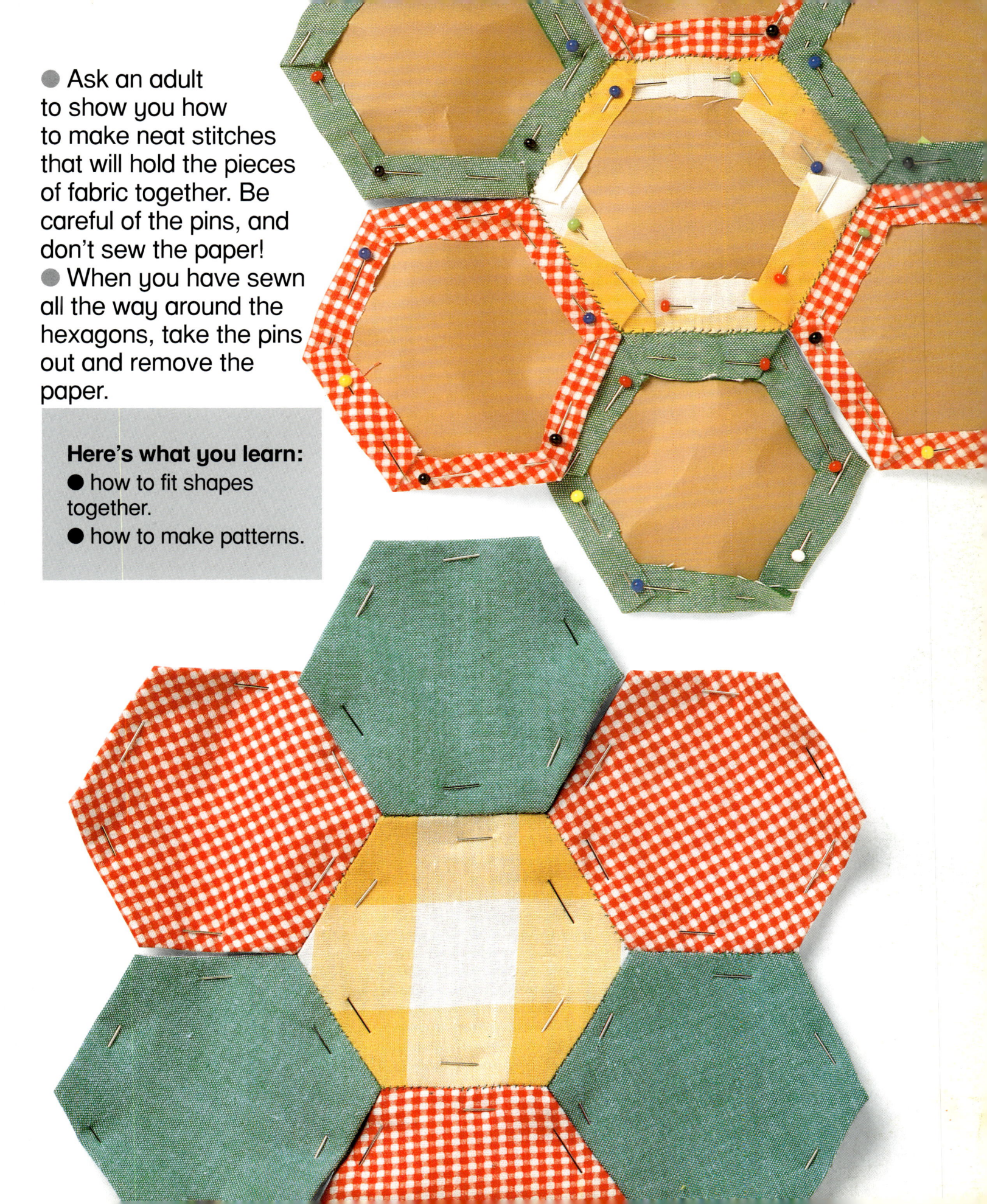

● Ask an adult to show you how to make neat stitches that will hold the pieces of fabric together. Be careful of the pins, and don't sew the paper!

● When you have sewn all the way around the hexagons, take the pins out and remove the paper.

Here's what you learn:
● how to fit shapes together.
● how to make patterns.

16 Cutout Shapes

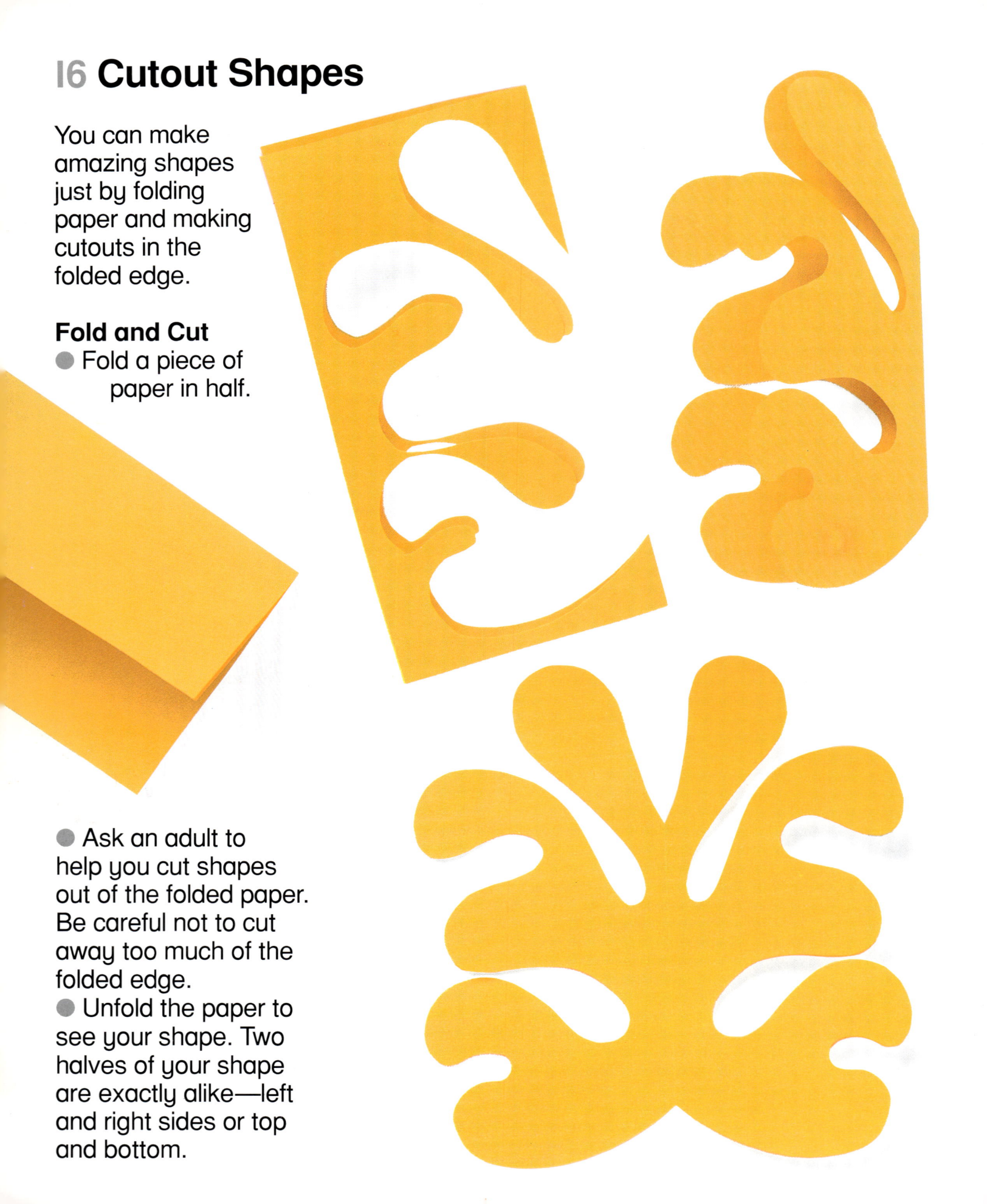

You can make amazing shapes just by folding paper and making cutouts in the folded edge.

Fold and Cut
- Fold a piece of paper in half.

- Ask an adult to help you cut shapes out of the folded paper. Be careful not to cut away too much of the folded edge.
- Unfold the paper to see your shape. Two halves of your shape are exactly alike—left and right sides or top and bottom.

Four Folds

- Now fold another piece of paper in half, and then in half again.
- Cut shapes from both folded edges, but remember not to cut off too much!
- When you unfold the paper you will find that the right and left sides match, and the top and bottom match, too.

Here's what you learn:
- about symmetry.

18 Slit-and-Slot Shapes

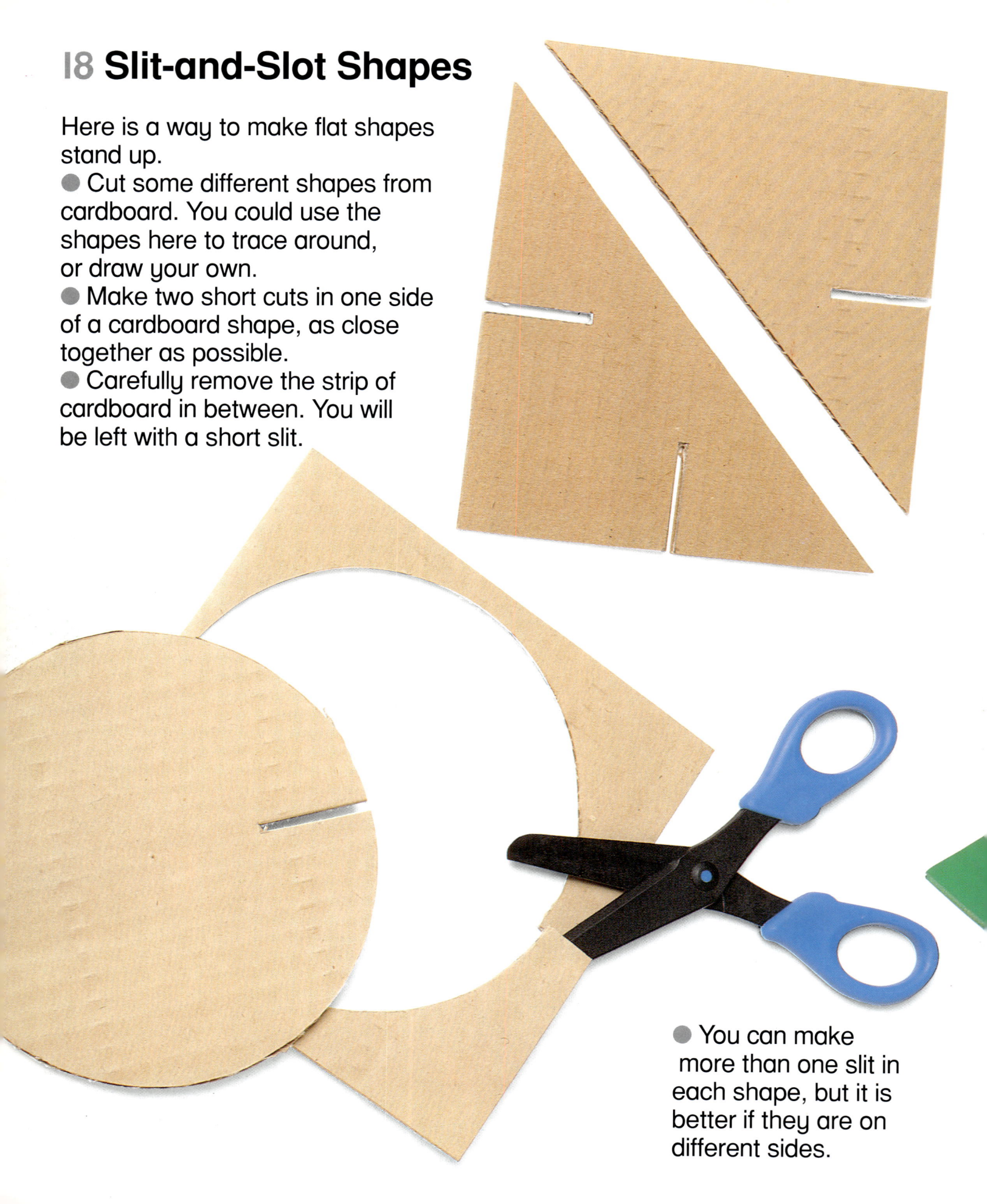

Here is a way to make flat shapes stand up.
- Cut some different shapes from cardboard. You could use the shapes here to trace around, or draw your own.
- Make two short cuts in one side of a cardboard shape, as close together as possible.
- Carefully remove the strip of cardboard in between. You will be left with a short slit.

- You can make more than one slit in each shape, but it is better if they are on different sides.

● How many different models can you make with your cardboard pieces? Try to make one that will stand up easily. Then, take the pieces apart and start again. What can you make this time?

● Paint the shapes different colors and let them dry. Then slot them together by pushing one slit into another.

Here's what you learn:
● about links between two-dimensional and three-dimensional shapes.

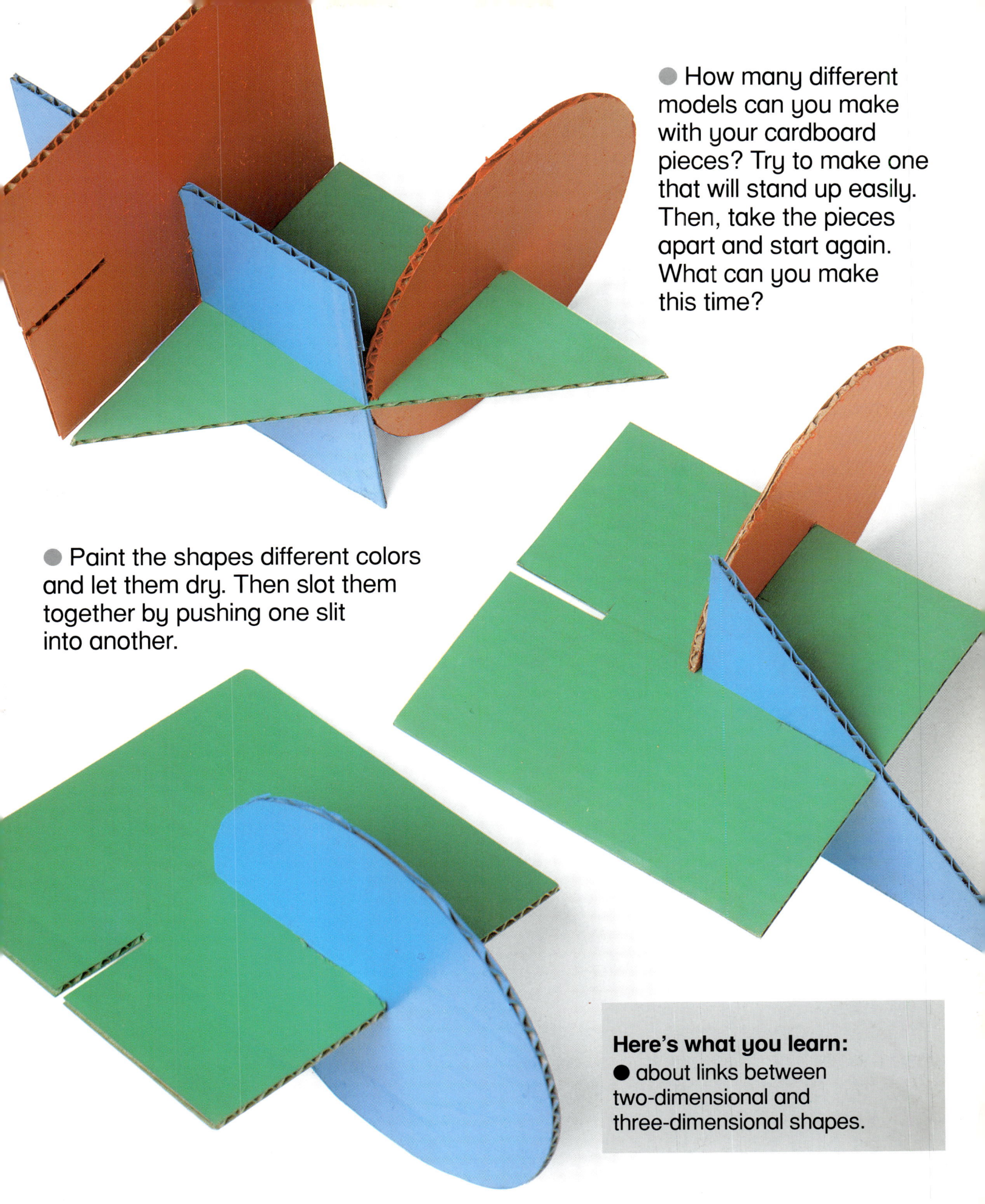

20 Stand-Up Jungle

Once you know how to slit and slot, you can make all sorts of models that stand up by themselves. Use thick paper or thin cardboard to make your models.

Slit-and-Slot Bushes
● Draw two bush shapes roughly the same size and cut them out. Cut a slit from the bottom of one shape to the middle. Then, cut a slit from the top of the other shape to the middle. Paint the shapes bright colors. Slot them together and stand the bush up.

Tall Trees
● Draw two tree shapes. You could use the tree here as a guide to trace around. Cut the shapes out. Make slits in both trees, as you did for the bushes. Now stand the tree up.

Palm Trees
● For these trees, cut out some large, leafy branch shapes. Cut a slit halfway across each branch near one end. Make slits into the top of a cardboard tube and slot in the branch.

Stand-Up Crocodile
● Cut out the shape of a crocodile's body. Then cut two leg shapes like the ones shown. Paint all the pieces. Make two slits in the base of the body and in the top of the leg pieces. Slot the legs into the body. Stand your crocodile up!

Now arrange your trees and bushes to make a jungle scene. You can make tall or short trees by using different sized cardboard tubes.

Hanging Bird
● Make a bird to keep the crocodile company. You will need three pieces — one for the body, one for the wings, and one for the head.

Here's what you learn:
● about three-dimensional shapes.
● about symmetrical shapes.
● about matching.

● Try making other kinds of plants. Can you think of any other slit-and-slot animals to put in your jungle?

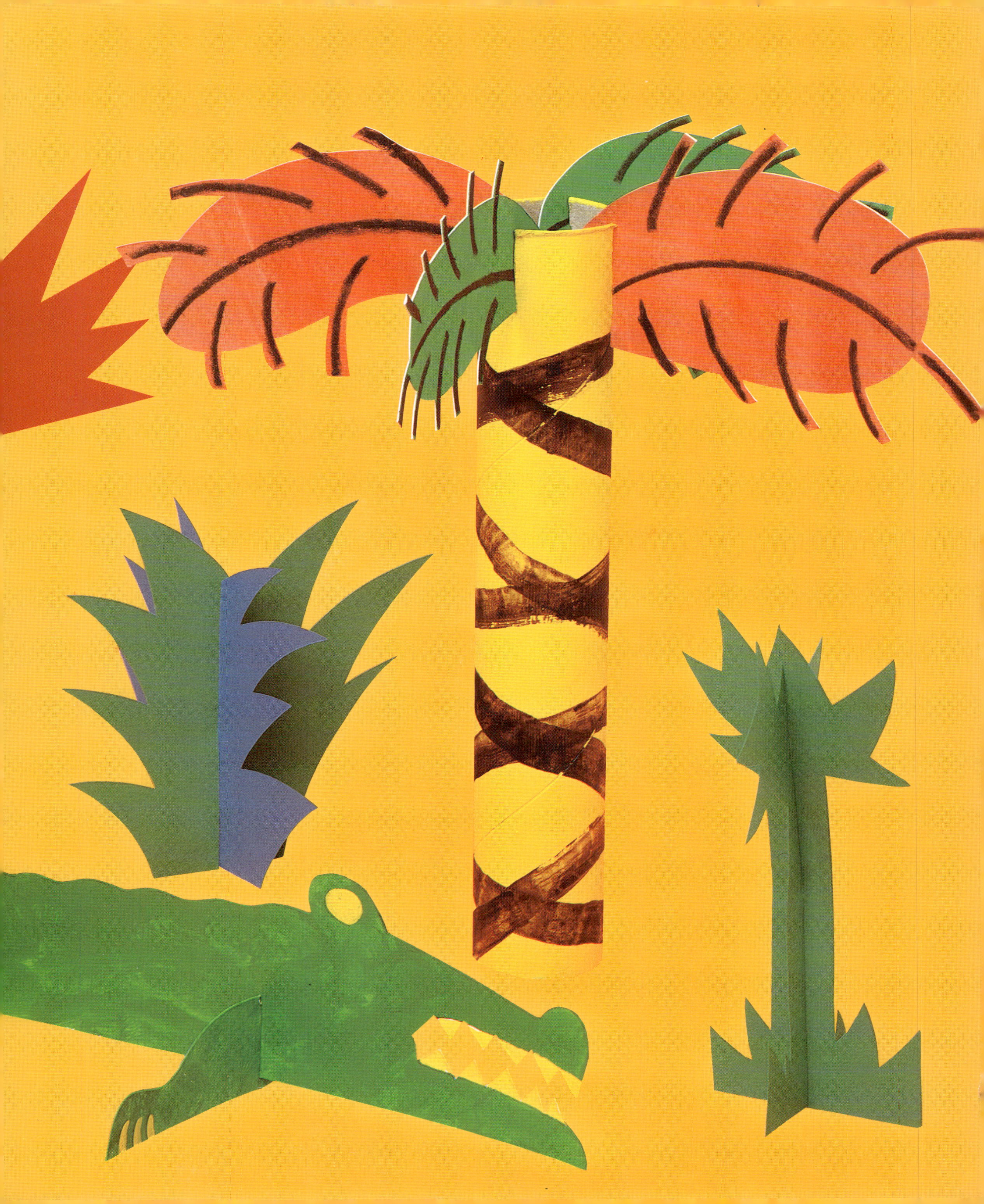

24 Food Monster

Look at all the different shapes of vegetables on this page. Some of them are very strange! Can you think of a way to describe them? Try making a funny food monster to decorate your table. Use round shapes to make the eyes and nose, and long shapes for the arms and legs.

- Ask a grown-up to help you join the vegetables together with toothpicks.
- Push one end of a toothpick into a vegetable.
- Then, push another vegetable onto the other end of the toothpick.

● You could also experiment by cutting fruits and vegetables into different shapes. Always ask an adult to help when you are using a knife.

Here's what you learn:
● how to make things using three-dimensional shapes.

26 Looking at 3-D Shapes

Containers come in lots of shapes and sizes. The next time you are in a store, see how many different shapes you can spot.

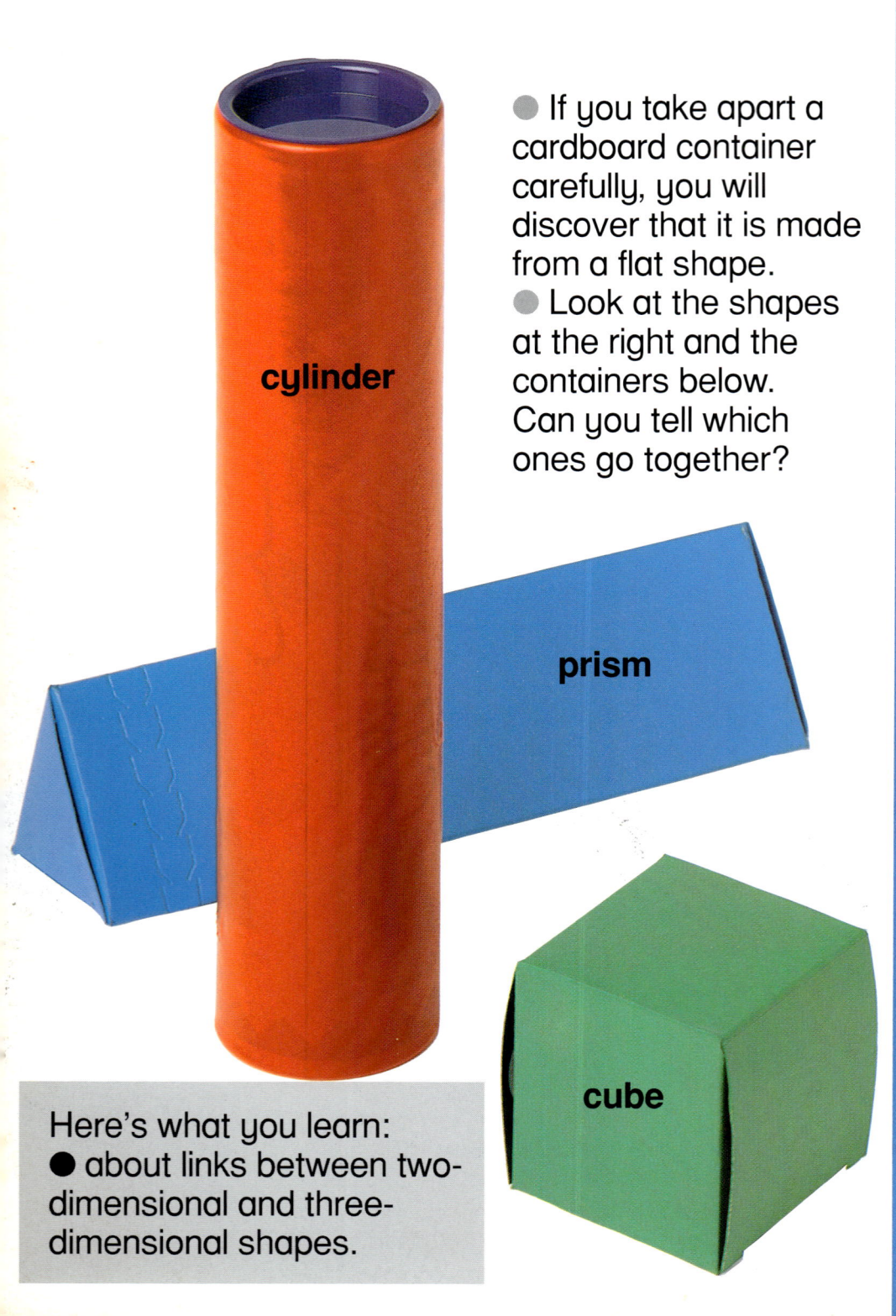

- If you take apart a cardboard container carefully, you will discover that it is made from a flat shape.
- Look at the shapes at the right and the containers below. Can you tell which ones go together?

cylinder

prism

cube

Here's what you learn:
- about links between two-dimensional and three-dimensional shapes.

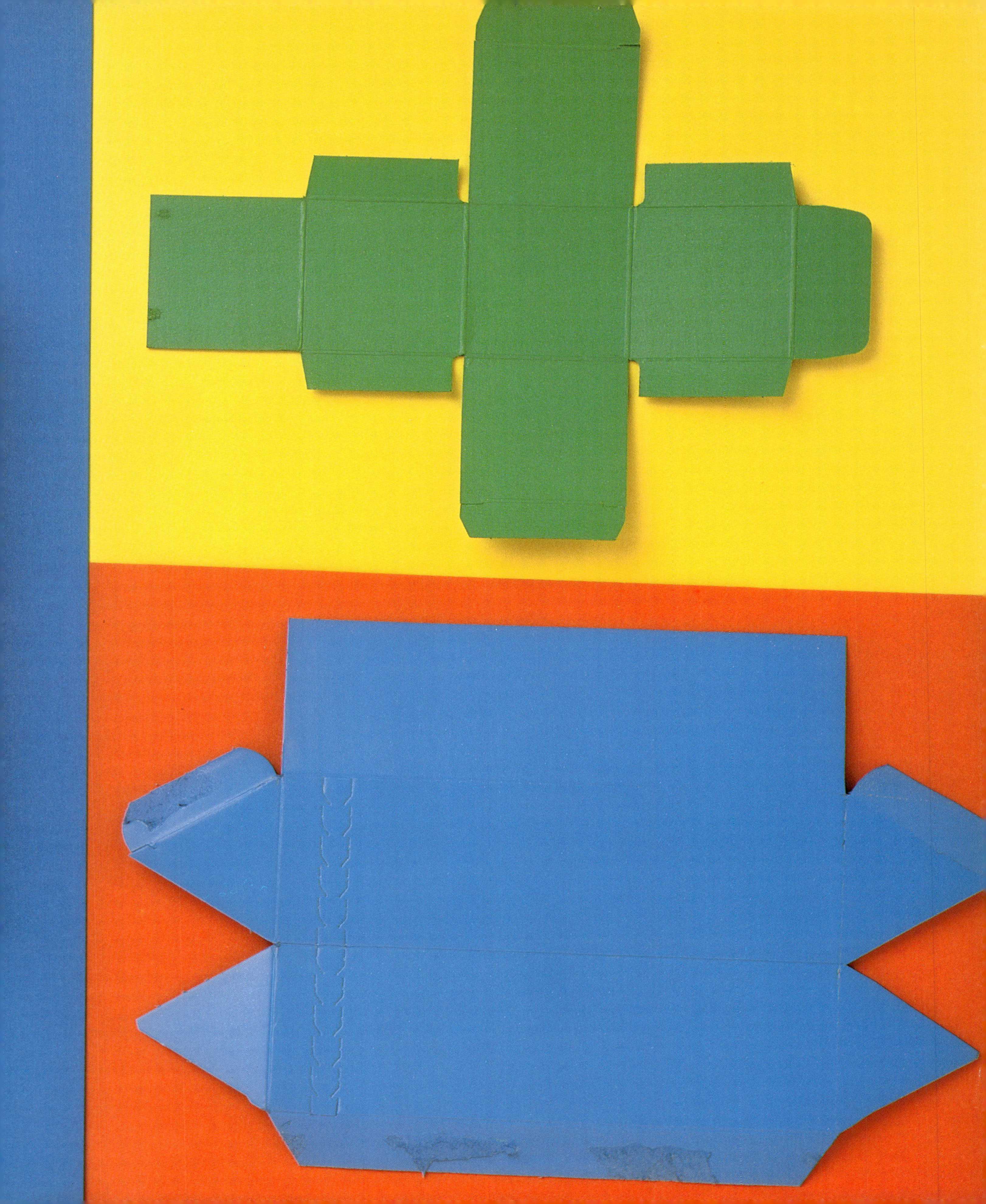

28 Make a Cube

On pages 26 and 27 are some containers that are opened out. If you follow the instructions here, you will be able to turn a flat piece of cardboard into a cube.
- Trace the shape on this page. Use your tracing to cut out the same shape from a piece of thin cardboard.

- Use a pencil and ruler to draw in the dotted lines, using the picture as a guide.

- You need to make the shape as precisely as possible so that all the sides of your cube will fit together.

The three narrow strips with shaped corners are called tabs.
- Fold the cardboard along each of the dotted lines, keeping the pencil marks on the inside of the cube.

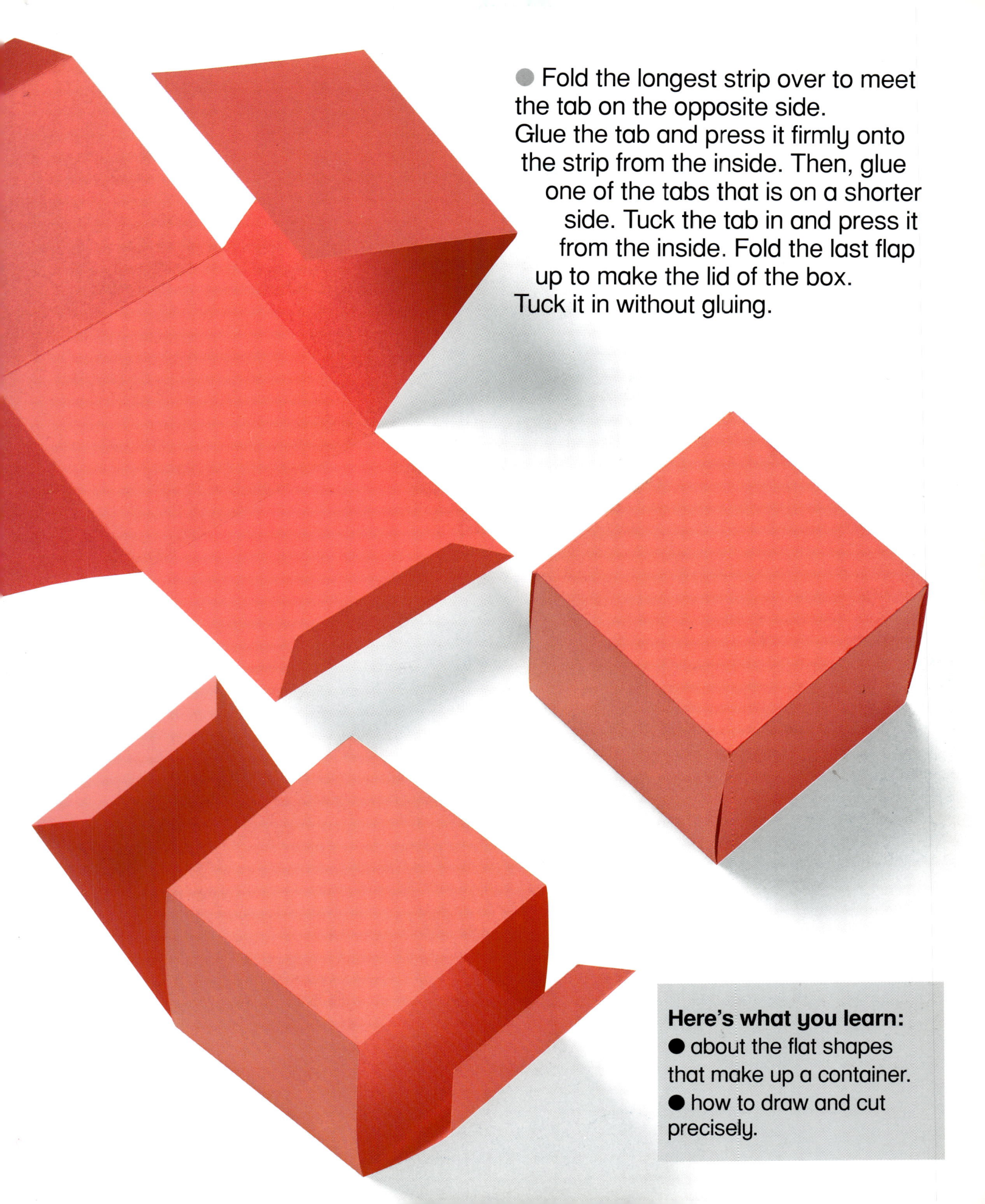

● Fold the longest strip over to meet the tab on the opposite side. Glue the tab and press it firmly onto the strip from the inside. Then, glue one of the tabs that is on a shorter side. Tuck the tab in and press it from the inside. Fold the last flap up to make the lid of the box. Tuck it in without gluing.

Here's what you learn:
● about the flat shapes that make up a container.
● how to draw and cut precisely.

30 Desk Organizer

You should be able to find cardboard tubes in lots of different shapes and sizes. They are often used for packaging, or inside rolls of paper and aluminum foil. Collect some tubes and turn them into a useful storage unit for your pencils, scissors, and erasers.

● Arrange your painted tubes in a group and glue them together along the sides.

Sorting
● First, decide which tubes are best for the things you want to store in them. For example, your pencils may need a long, thin tube. If necessary, you could ask an adult to cut a tube to make it shorter.
● Paint the tubes bright colors.

● Make a base for your storage unit by gluing the bases of the tubes to a piece of cardboard. Fill the tubes with all your pencils, pens, and crayons.

Here's what you learn:
● how to recognize shapes in different sizes.

32 Build an Animal

Our colorful animal was made from used boxes, a ball, and tubes. Start collecting as many shapes as you can. Get permission to use the things you collect. Look for interesting shapes with curved and straight edges. If you are using containers that have had food in them, make sure you clean them first.

● You can use sections cut from boxes to make different shapes. We made the triangle out of the corner of a box and the round shape (a sphere) is a painted foam ball.

- Sort your collection of boxes and tubes into different shapes and sizes.
- Plan what you are going to build.
- Decorate the boxes and tubes with different colors and patterns.

Here's what you learn:
- about three-dimensional shapes.
- how to make one shape into another.

- Use your boxes, cartons, and tubes to build with. You could glue or tape them together to keep them in position, or just balance them carefully.

34 Stretchy Jewelry

Here is a way to turn flat pieces of paper into completely different shapes that you can wear. Your friends will be amazed! You will need long strips of paper in two different colors.

Accordion Folds
- Glue the ends of two strips of paper together at right angles. Look at the picture at the top.
- Then, fold the strip that is underneath over the top strip.
- Next, fold the strip which is now underneath over the top one.
- Continue folding until you get to the ends of the strips.
- Glue the ends down.

- Gently pull the two ends of the paper to open the chain a little. When you let go, the shape will spring back. To make a longer piece, glue extra strips of paper onto the end of the chain.

You can use your stretchy paper chains to make colorful necklaces, earrings, and bracelets.

Here's what you learn:
- how to turn a two-dimensional shape into a three-dimensional one.
- how to make right angles.

Measure

38 What Size Is It?

Every day we ask questions such as:
How big is it? How tall is it?
How heavy is it?
What time is it?
All of these questions have something to do with measuring. In this book we will be looking at lots of different ways of measuring.

We use words such as big and small, tall and short, or heavy and light to describe things around us.

Look at the groups of objects on these pages. Which is the biggest in each group? Which is the smallest?

Measuring is one of the most important math skills we need to learn.
The activities in this book will help you:
- explore height, length, weight, area, and volume.
- use measures.

40 How Tall Are You?

Make your own height chart so that you can measure yourself.
- Glue or tape several large pieces of colored paper to make one long strip that is taller than you are.
- Use a building block to make equally spaced lines down the side of the paper.

Colored Strips
- Use a ruler and the block to measure strips of colored paper. The strips should be the width of your chart and the height of your block.
- Cut out the strips carefully.
- Glue one strip along the bottom of your height chart. Leave a space and glue another strip between the second and third lines from the bottom. Keep gluing equally spaced strips until you reach the top.

Here's what you learn:
- how to make and use measurement.

- Hang the chart on a wall so that the bottom touches the floor.

● Stand next to your chart and ask someone to make a mark on it right over your head. Count the number of strips from the floor up to the mark. Now measure your friends against your chart. Who is the tallest? Who is the shortest?

● Measure yourself and your friends again in one month. Has anyone grown taller?

42 Puppet Play

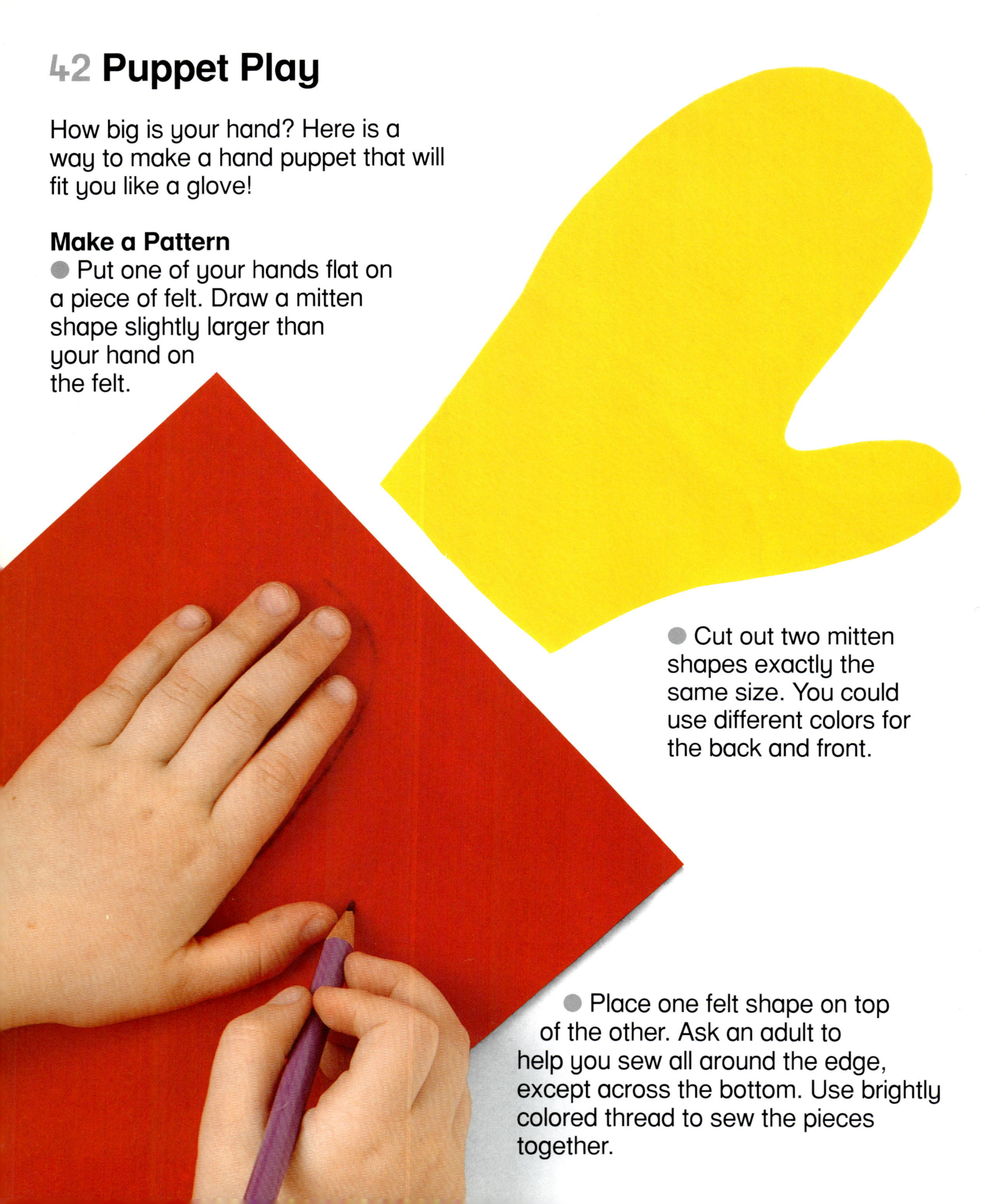

How big is your hand? Here is a way to make a hand puppet that will fit you like a glove!

Make a Pattern
- Put one of your hands flat on a piece of felt. Draw a mitten shape slightly larger than your hand on the felt.

- Cut out two mitten shapes exactly the same size. You could use different colors for the back and front.

- Place one felt shape on top of the other. Ask an adult to help you sew all around the edge, except across the bottom. Use brightly colored thread to sew the pieces together.

● Now you are ready to decorate your puppet. Can you guess what these puppets are going to be? Turn the page to find out.

Here's what you learn:
● how to investigate area.
● how to compare sizes.
● how to match shapes.

Here are the finished puppets! Just by adding a few bits and pieces you can turn your puppets into real characters. You can buy eyes to glue on or make your own from felt. Ears, noses, mouths, and tusks can also be cut from felt and glued in place. Why not invent your own puppet?

Frog
● Use green felt to make a frog puppet. Then give it a wide, red mouth.

Monster
● Cover your mitten with spots to make a funny monster puppet.

Elephant
● Use gray felt to make ears and a trunk for an elephant. Then add a pair of tusks to finish it off.

Shape Face
● Make a face using squares for eyes, a triangle for the nose, and a long, thin rectangle for the mouth. This puppet looks a little like a robot!

46 Finger Puppets

Here is an easy way to make a puppet that will fit on your finger.

Bird
- Cut out a small paper circle. Make a cut from the edge into the center.
- Fold the paper around to make a cone and tape the edge.
- Tape the cone onto the paper tube.
- Cover the paper tube and cone with several layers of paste and pieces of torn newspaper. You can make your own paste by gradually mixing water and flour until the mixture is thick and creamy.

Making a Tube
- Take a long strip of paper or thin cardboard and wrap it snugly around your finger.
- Tape the end and then gently pull the tube off your finger.

● When your puppet is dry, you can paint it. Why not make a puppet to fit every finger?

Sombrero
● Make a tube that fits your finger, as you did before.
● To make the sombrero, cut out a paper circle. Ask an adult to help you make some small slits in the center of the circle. Push the circle over the paper tube to make the brim of the hat.
● Cover the tube and brim with newspaper and paste as before.

Here's what you learn:
● how to investigate size.

48 Cutout Mask

We used a paper plate to make our mask. You will need to know where to make holes for your eyes, nose, and mouth. Here is a good way to take the measurements.

Measure It Out
● To measure the distance between your eyes, hold a piece of yarn in front of your face, stretching from the middle of one eye to the middle of the other. Then, lay the yarn in the middle of your mask and mark each end.

● Next, measure the length of your nose. Stretch a piece of yarn from the middle of your eyes to the bottom of your nose. Use the yarn to mark the mask.

● Finally, stretch a piece of yarn from the bottom of your nose to the middle of your mouth and mark it on the mask.

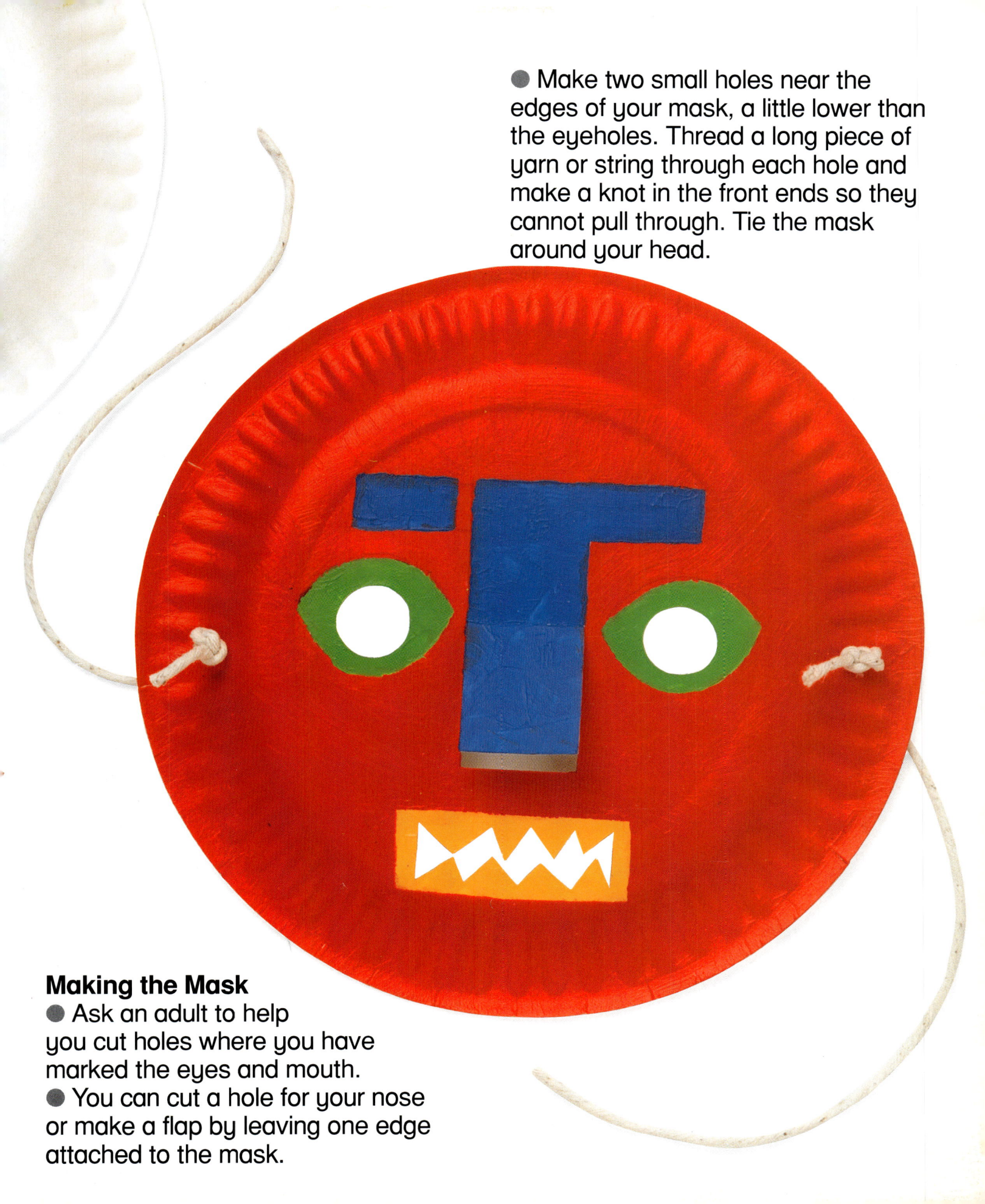

● Make two small holes near the edges of your mask, a little lower than the eyeholes. Thread a long piece of yarn or string through each hole and make a knot in the front ends so they cannot pull through. Tie the mask around your head.

Making the Mask
● Ask an adult to help you cut holes where you have marked the eyes and mouth.
● You can cut a hole for your nose or make a flap by leaving one edge attached to the mask.

Once you know how to make a basic mask, you can make up all sorts of different ways to decorate it.

Tiger
● First, paint a striped tiger face on your mask.

● Next, cut out some paper ears, paint them, and glue or tape them to the top of the mask.

Bird
The feathers on this bird mask are made from paper.
● Cut feathers of different sizes.

● Glue the large feathers on first and the smaller ones on top of them. To make the beak, cut a piece of folded paper. Unfold it and glue it in place.

Here's what you learn:
● how to use measurements.

52 Paper Dolls

Here is a way to make a cardboard doll and a whole wardrobe of paper clothes. First, draw the outline of a person on cardboard or trace around the doll shown below. Ask an adult to help you cut it out. Then, make some clothes…

Making a Shirt
- Place the doll on colored paper and draw around the top half of the body.
- Lift the doll off and draw a shirt shape slightly larger than the doll's body.
- Draw two tabs on the top of the shirt at the shoulders. Ask an adult to help you cut out the shirt. Put the shirt on the doll and bend the tabs over to hold it in place.

Making a Hat
- Draw a hat shape a little bigger than the doll's head.
- Ask an adult to cut a slit in the hat like the one shown below. Push the doll's head through the slit.

Here's what you learn:
- how to match sizes and shapes.
- how to investigate area.

Making Pants
- Place the doll on a different color of paper and draw around the bottom half of it.
- Lift the doll off and draw a pair of pants slightly larger than the doll's legs. Add some tabs at either side of the waist.

54 Bottle Band

It is hard to believe, but you really can make your own band with some empty glass bottles, water, a metal spoon, and a plastic cup to use for measuring.

- Make a mark about a quarter of the way up an empty plastic cup.
- Fill the cup with water up to the mark. We added a little food coloring to the water to brighten up our bottle band.
- Pour the water from the cup into the first bottle. You may need a pitcher or a funnel to help.

- Pour two measures from the cup into the second bottle.

● Now pour three measures into the third bottle, four into the fourth bottle, and five into the fifth bottle.

● Play your bottle band by knocking gently on the sides of the bottles with a metal spoon.

Here's what you learn:
● how to measure using units.

56 A Mobile

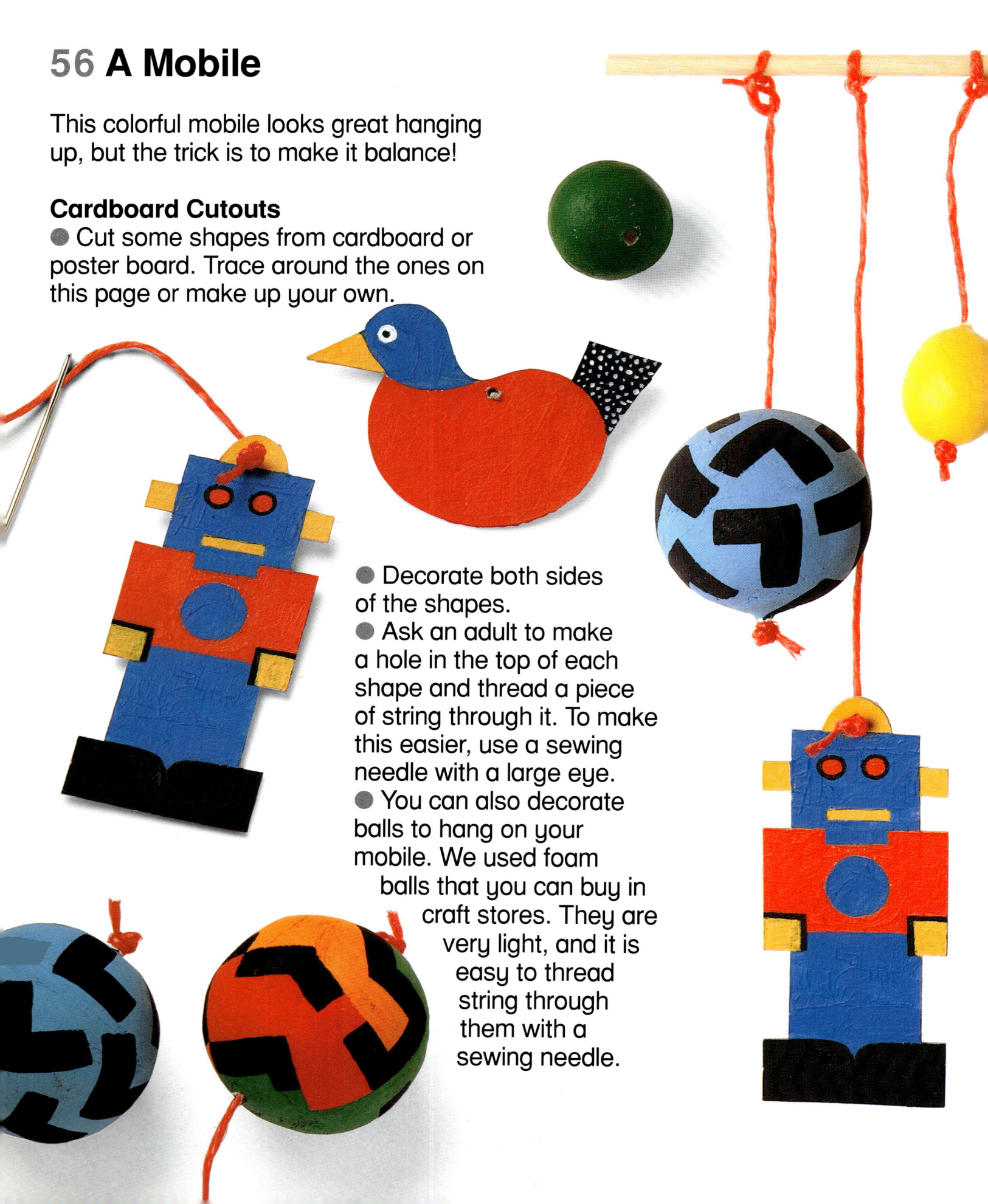

This colorful mobile looks great hanging up, but the trick is to make it balance!

Cardboard Cutouts
- Cut some shapes from cardboard or poster board. Trace around the ones on this page or make up your own.

- Decorate both sides of the shapes.
- Ask an adult to make a hole in the top of each shape and thread a piece of string through it. To make this easier, use a sewing needle with a large eye.
- You can also decorate balls to hang on your mobile. We used foam balls that you can buy in craft stores. They are very light, and it is easy to thread string through them with a sewing needle.

Hang It Up
● Attach a piece of string to the middle of a long stick or dowel. Use this string to hang your mobile.

● Tie the objects to the stick in different places. Slide them along the stick until the mobile balances. You may also have to adjust the lengths of the strings so that the objects don't touch each other.

Here's what you learn:
● about weight and balance.

58 Baking Cookies

It's very important to measure carefully when you are baking. Ask an adult to help you measure the right amount of each ingredient.

You Will Need:
3/4 cup all-purpose flour
1/4 cup granulated sugar
1/3 cup butter
2 drops vanilla extract

● Ask an adult to heat the oven to 325°F.

Mix It Up
- Put the flour and sugar into a mixing bowl.
- Cut the butter into small pieces and add it to the bowl. Use your fingertips to rub the butter into the flour. (Make sure your hands are clean!)
- When the mixture looks like fine bread crumbs, add the vanilla extract. Use your hands to make the mixture into a large ball of dough.

Roll It Out
- Press the dough on a surface lightly sprinkled with flour.
- Use a rolling pin to roll the dough into a thin, flat shape that is about the same thickness as a coin.
- Cut shapes using a cookie cutter.
- Put the cookies on a greased cookie sheet and bake them for 20 minutes.
- Ask an adult to take the cookies out of the oven. Use a spatula to put the cookies on a wire rack to cool.

Here's what you learn:
- how to use standard measures.

60 Cookie Boxes

When you have made your cookies, you might want to give some away as a present. Here are two ways of making boxes in which to pack them.

Pile Them Up

● Put one of the cookies on a piece of cardboard. Using a ruler, draw a square slightly bigger than the cookie. Cut out two squares exactly the same size. These will be the top and bottom of the box.

● Decide how many cookies you want to give away and pile them up. Stand a piece of cardboard against the pile to measure the height. Make a mark on the cardboard.

● Use a ruler to draw a cardboard rectangle slightly taller than the pile of cookies and the same width as the square you have already cut out. Cut four rectangles this size. These will be the sides of the box.

● Tape the bottom and sides together. Put the cookies in the box and tape the last square on top to make the lid.

Spread Them Out

- Lay six cookies flat on a piece of cardboard and draw a rectangle around them. Cut out two rectangles that size.
- Cut four thin strips of cardboard—two the same length as the rectangle and two the same width as it.
- Tape the strips to the rectangle to make the sides of the box.
- Tape the other rectangle on top to make the lid.

Here's what you learn:
- about area and volume.

62 Wrapping Boxes

Make some fun paper to wrap around your box by painting some plain paper.

Paint and Paper
- First, choose a piece of paper that will be big enough to cover the box.
- Splatter some paint onto the paper using a brush, or dab it on with a sponge dipped in paint.

Wrap It Up
- When the paint is dry, lay the paper with the decorated side down and place the box in the middle. Fold the long sides of the paper around the box.

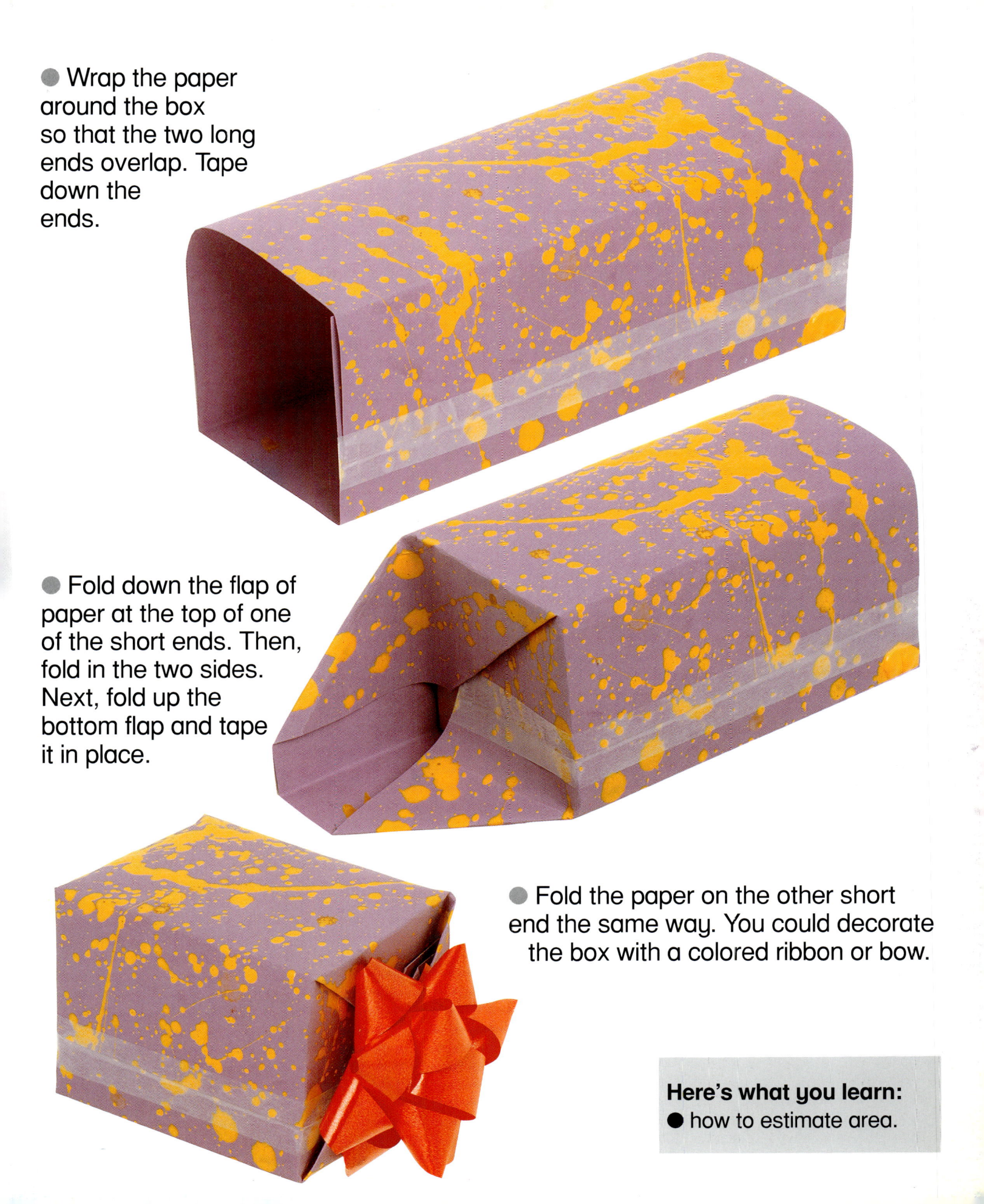

● Wrap the paper around the box so that the two long ends overlap. Tape down the ends.

● Fold down the flap of paper at the top of one of the short ends. Then, fold in the two sides. Next, fold up the bottom flap and tape it in place.

● Fold the paper on the other short end the same way. You could decorate the box with a colored ribbon or bow.

Here's what you learn:
● how to estimate area.

64 Sand Timer

Here's a way to measure how much time something takes.

Paper Cone
● Draw a big circle on a piece of poster board. You could trace around a big plate.

● Cut out the circle. Then, ask an adult to help you cut a slit from the edge of the circle to the center.

Here's what you learn:
● how to measure time.

- Fold one side of the slit around the other to form a cone shape, like the one below. Tape it in place.
- Cut a tiny hole in the bottom of the cone.

A Line of Triangles
- Cut out several paper triangles all the same size. Glue them in a line up the side of an empty bottle.

Falling Sand
- Place the cone in the top of the bottle and fill it with sand. Watch the sand fall through the cone and fill up the bottle. The sand gradually reaches each triangle mark.
- Have a friend watch your timer while you hop around the room. How many marks does the sand pass? Does your friend take the same amount of time to hop? What else can you time?

66 Tangrams

A tangram is a square that has been cut into seven special pieces. The pieces can be put together to make different patterns or pictures. Look at the pictures shown on this page. Can you find a small square and five triangles in each picture? The black shape is called a parallelogram. It is also in all of the pictures.

Making a Tangram
- Ask an adult to cut out a square from thick cardboard.
- Use a ruler and pencil to mark the shapes shown above.

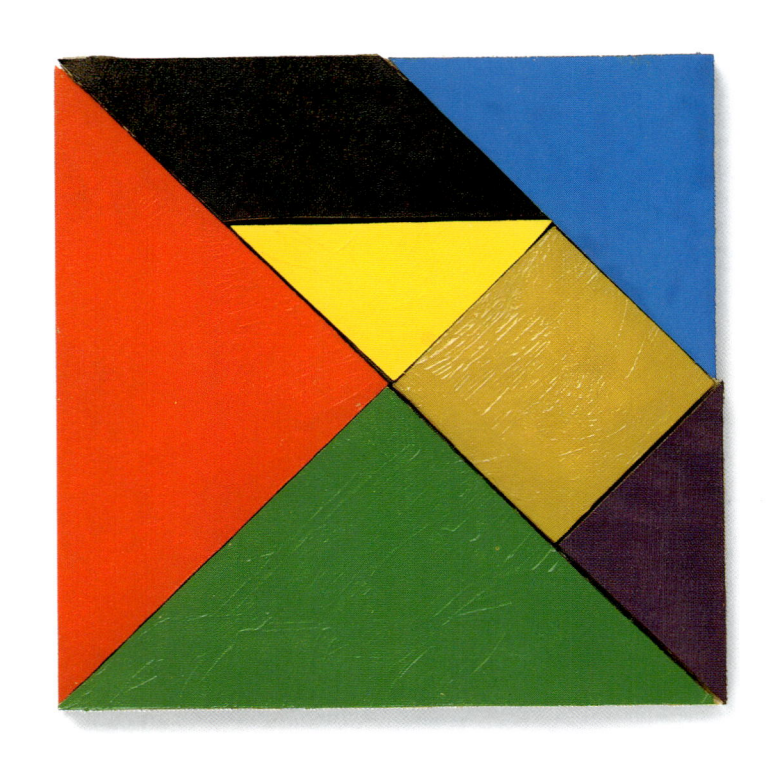

- Cut along the lines and paint each piece a different color.

- How many pieces make up the tangram? Mix them all up. How many pieces do you count now?

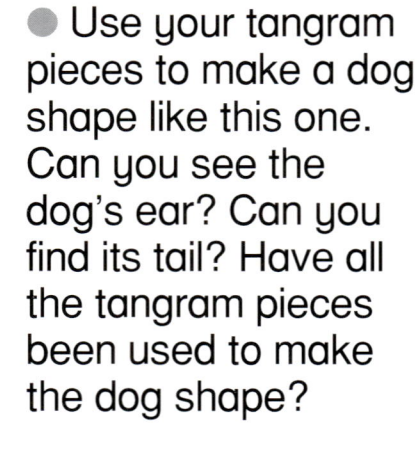

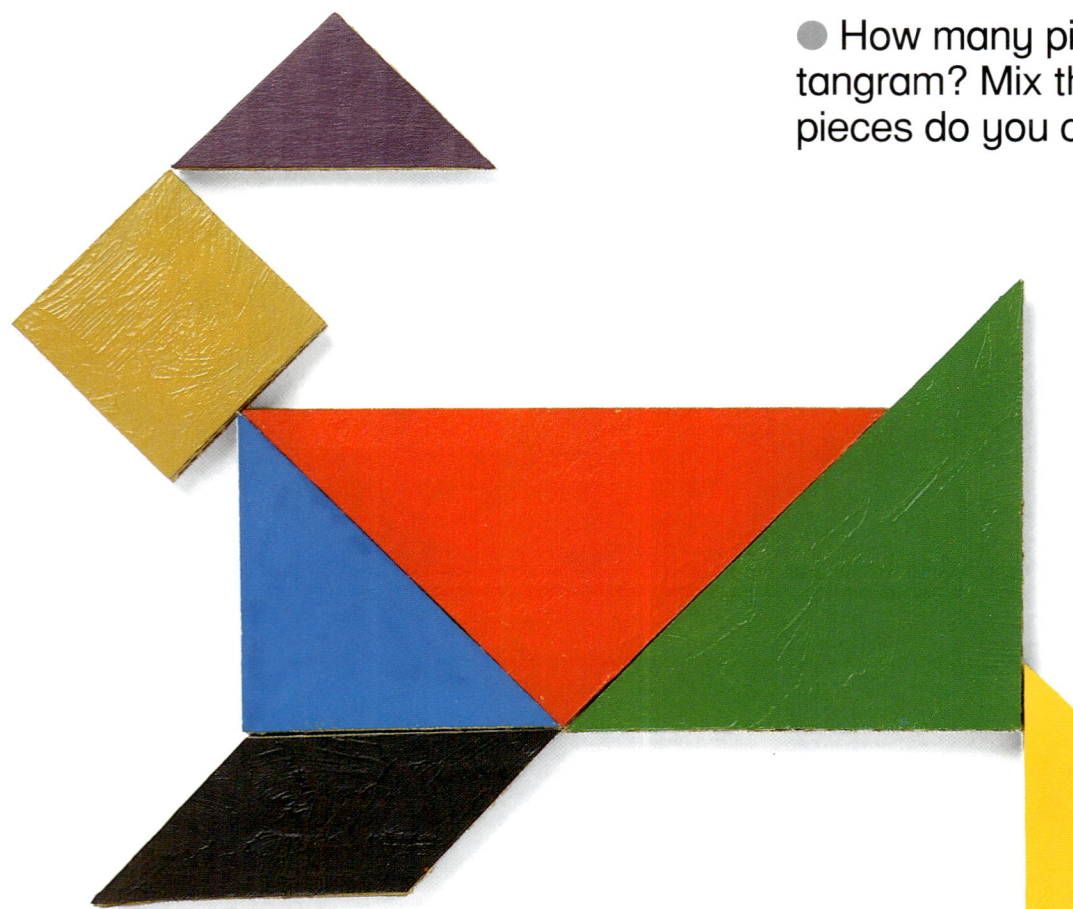

- Use your tangram pieces to make a dog shape like this one. Can you see the dog's ear? Can you find its tail? Have all the tangram pieces been used to make the dog shape?

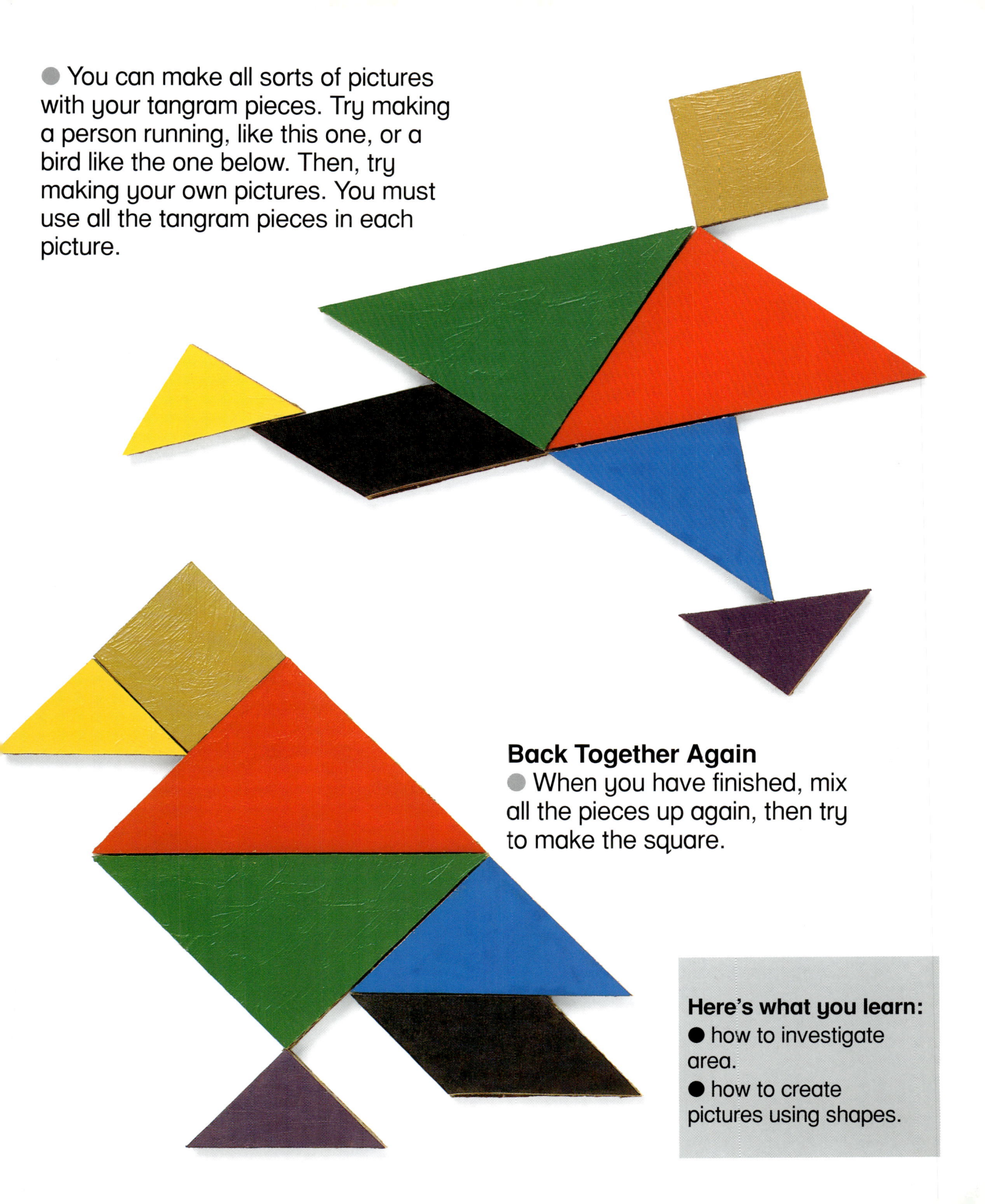

● You can make all sorts of pictures with your tangram pieces. Try making a person running, like this one, or a bird like the one below. Then, try making your own pictures. You must use all the tangram pieces in each picture.

Back Together Again
● When you have finished, mix all the pieces up again, then try to make the square.

Here's what you learn:
● how to investigate area.
● how to create pictures using shapes.

Patterns

70 Looking at Patterns

There are patterns all around us. You can find them in sidewalks, in the stitches of your sweater, and on a butterfly's wings. A pattern is made when shapes or numbers are put in a sequence and repeated. Look around you. How many different patterns can you see?

We use patterns to help us make sense of the world. Math is all about patterns. The activities in this book will help you:
- sort things into groups.
- match similar things.
- find out how things fit together.
- create patterns, shapes, and designs.

72 Spiral Snake

Make a patterned snake to hang from your ceiling.

Painting a Spiral
- Draw a spiral on a piece of poster board. Start from the edge of the board and gradually spiral in toward the center. You may need to draw a few spirals for practice first.
- Cut out the snake starting from the end of the spiral on the edge of the poster board.

● Paint a snake pattern on your spiral, or decorate it with colored paper.
● Ask an adult to string a piece of thread through the middle of the spiral. Now hang up your snake.

Here's what you learn:
● how to create repeating patterns.
● how to change a flat shape into a three-dimensional one.

74 Beads

Threading beads is a good way to make a pattern. Look around your house or school for things to use as beads. Here are some ideas for making your own beads.

Paper Beads

● Glue together two pieces of colored paper. Tear out a triangle shape. Roll the shape around a pencil and tape down the narrow end.

● Make a simple paper bead with a long strip of colored paper. Roll it around a pencil, then tape down the end. You could decorate the paper before making your beads.

Pasta Beads

● Pick pasta shapes with holes through the middle.
● Paint the shapes and let them dry.

Clay Beads
- Use the type of modeling clay that dries overnight to make these beads.
- Make small balls of clay and ask an adult to make a hole in each with a knitting needle or toothpick. Let the beads harden.

Sort Them Out
How many different types of beads have you made or collected?
- Sort the beads into different colors and shapes.

Here's what you learn:
- how to sort things into different groups or categories.
- how to match similar things.

Beads to Find
If you look, you should be able to find lots of things to use as beads. We used straws, plastic beads, and even peanuts. Can you think of anything else you could use?

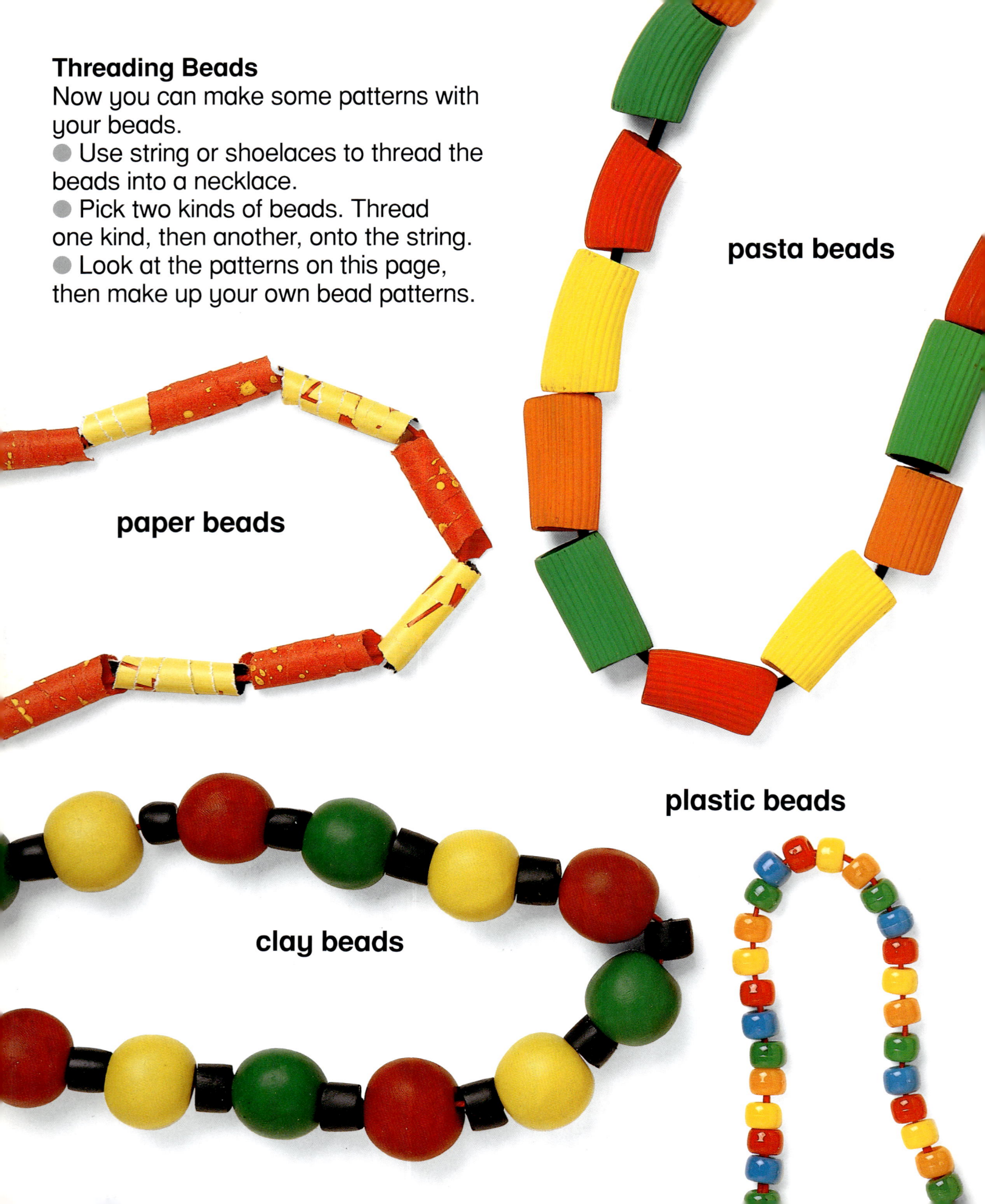

Threading Beads
Now you can make some patterns with your beads.
- Use string or shoelaces to thread the beads into a necklace.
- Pick two kinds of beads. Thread one kind, then another, onto the string.
- Look at the patterns on this page, then make up your own bead patterns.

pasta beads

paper beads

plastic beads

clay beads

paper beads

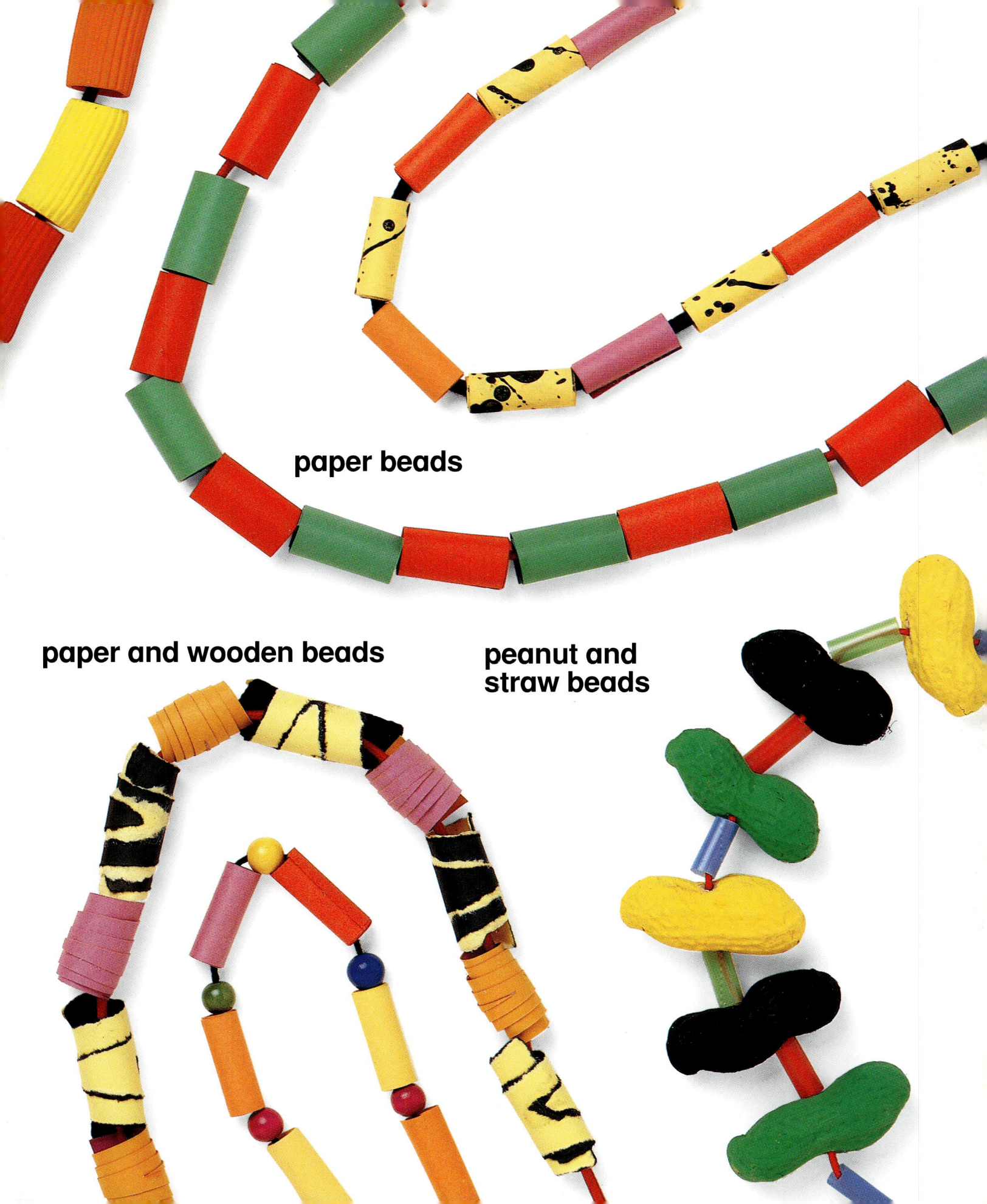

paper and wooden beads

peanut and straw beads

78 Cake Fun

Here's a delicious way to play with patterns! Ask an adult to help frost the top of a cake with icing. Then use different candies to decorate the top.

Shapes and Colors
First decide which sweets you are going to use. Which shapes and colors look good together? Which are your favorites?

Planning your Pattern

● Start by making a circle of different candies around the edge of the cake.

● Choose another pattern for the center of the cake. You could make more circles or use different candies to make a cross.

Here's what you learn:
● how to sort things into groups.
● how to match similar things.
● how to create simple patterns.

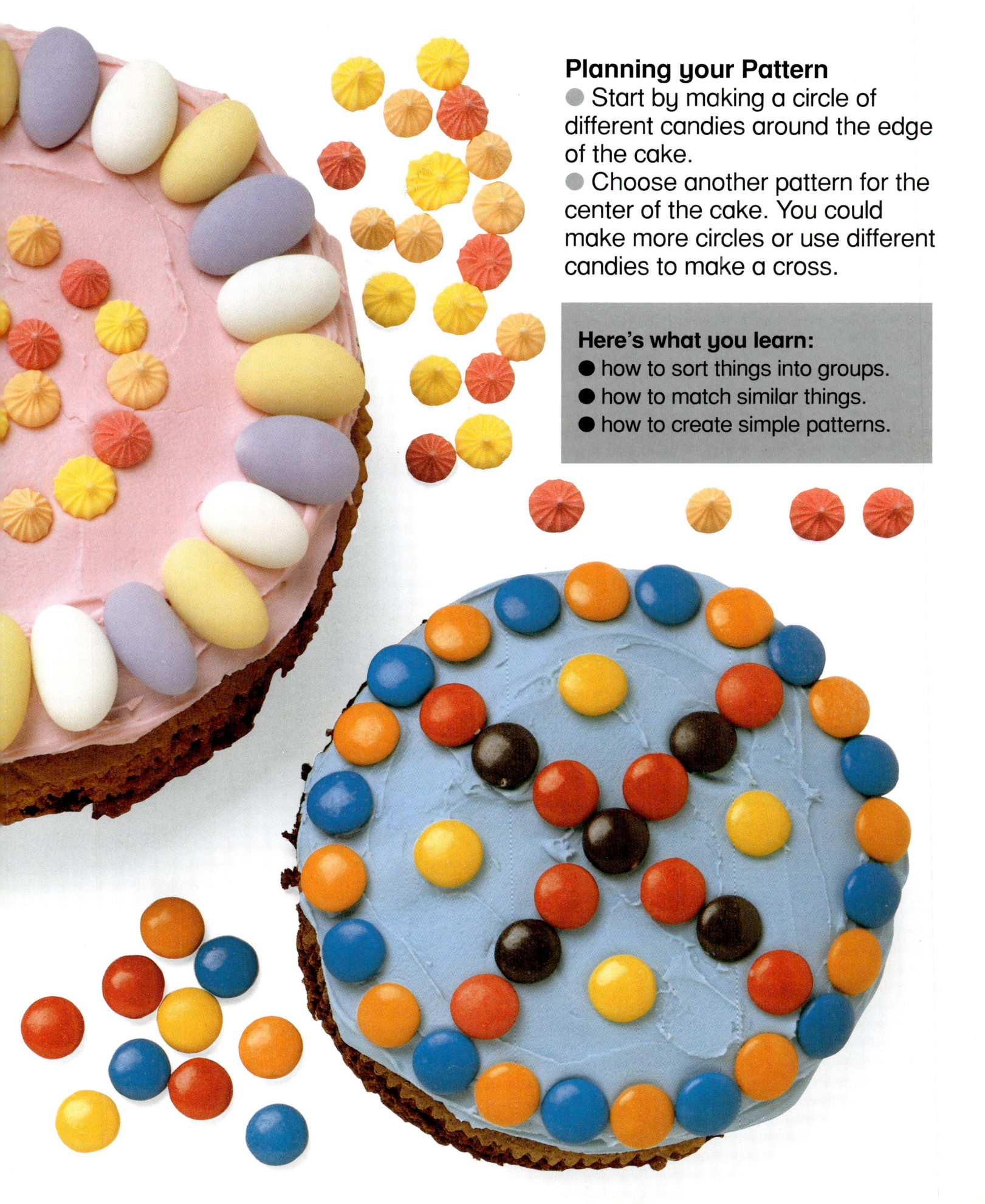

80 Weaving

Some clothes are made from woven fabrics. The threads make a pattern. These fabrics are made on large machines called looms. You can do your own weaving at home using a cardboard loom and brightly colored yarn or strips of felt.

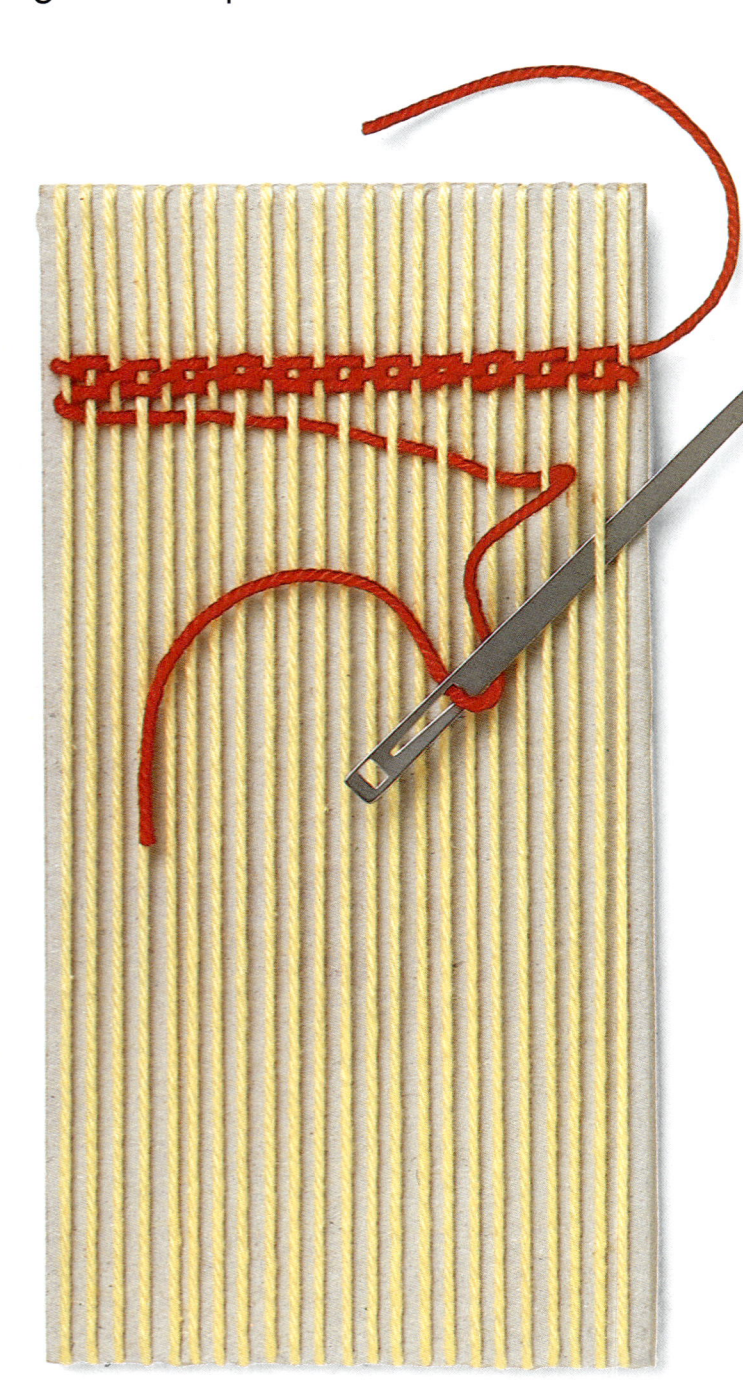

Loom
● Ask a grown-up to cut notches in two ends of a small piece of cardboard.
● Wind a piece of yarn around the cardboard. The notches will keep it in place. Tie the ends of the yarn at the back.

Over and Under
● Thread a large, blunt needle with a piece of yarn.
● Push the needle under and over the threads until you reach the other side.
● Weave back the other way, under the threads you went over before, and over those you went under.
● Cut the threads at the back to take your weaving off. Tie the ends together.

Here's what you learn:
● how to create simple patterns.
● how to use ideas about symmetry.

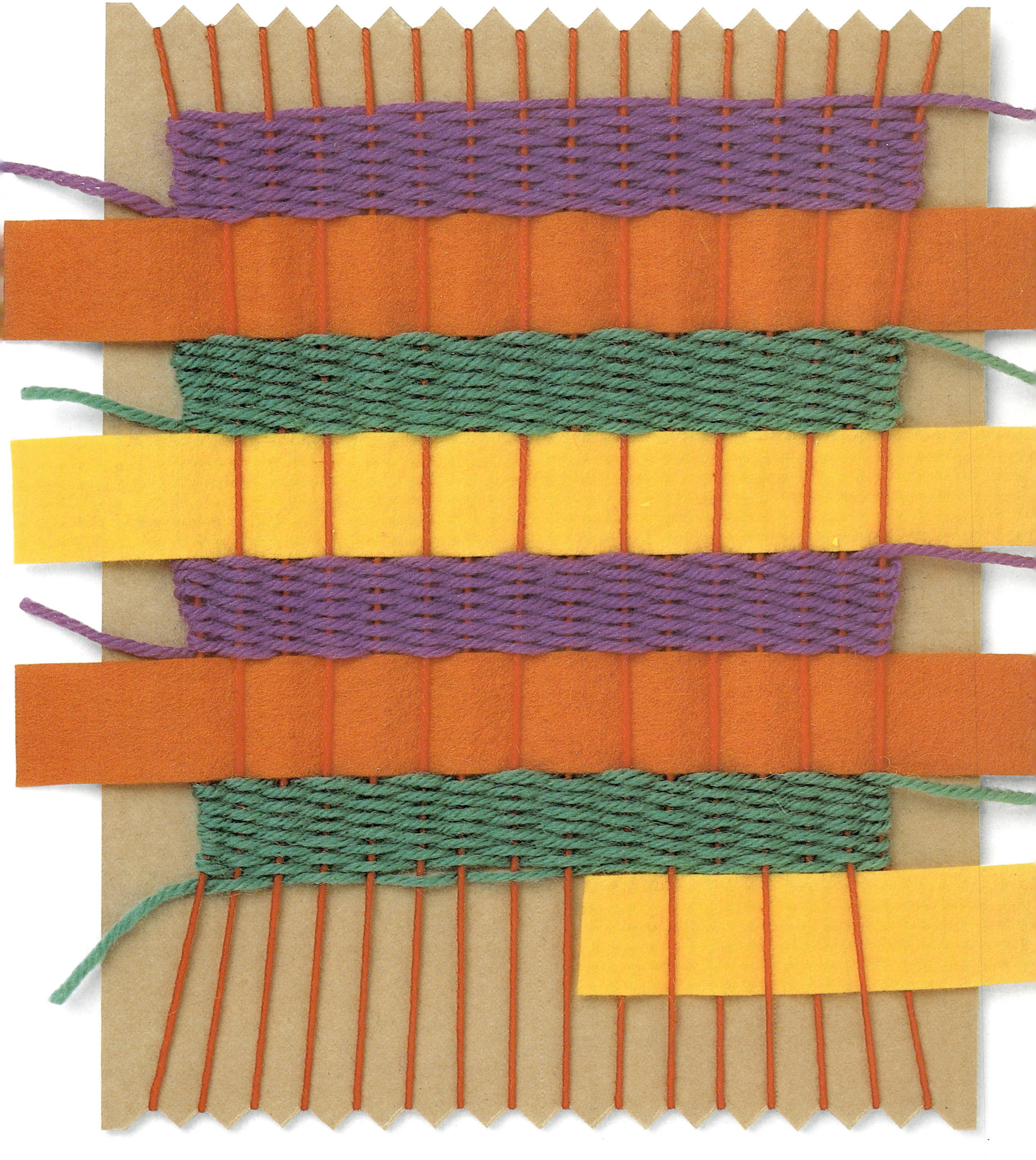

You can make a pattern with woven paper, too! First, find some fairly stiff colored paper.

Simple Pattern
- Fold a piece of paper in half.
- Make a row of cuts along the folded edge. Unfold the paper.
- Cut some strips of another color. Weave these strips through the slits as shown at right.

Diagonal Stripes
- Fold another piece of paper diagonally and make cuts. Weave strips as shown below or below right. Does the pattern look like the simple pattern?

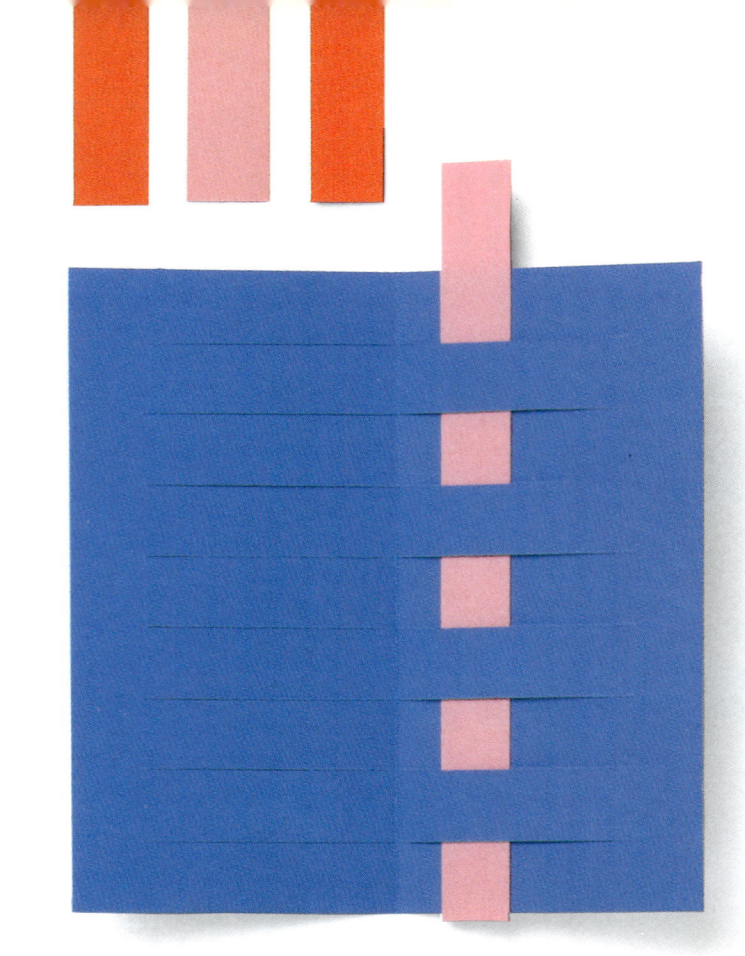

Here's what you learn:
- how to create patterns and shapes.

Wavy Stripes
Cut wavy slits in a piece of paper. You will also need some wavy strips of another color. Weave the strips as before.

Tartan Stripes
Cut two slits close together in a piece of folded paper. Leave a gap, then cut two more slits close together, and so on. Weave thick and thin strips of colored paper in and out of the slits.

Zig-zag Stripes
Ask an adult to make zig-zag slits in the paper with a craft knife. Weave straight strips through the slits.

84 Dot Patterns

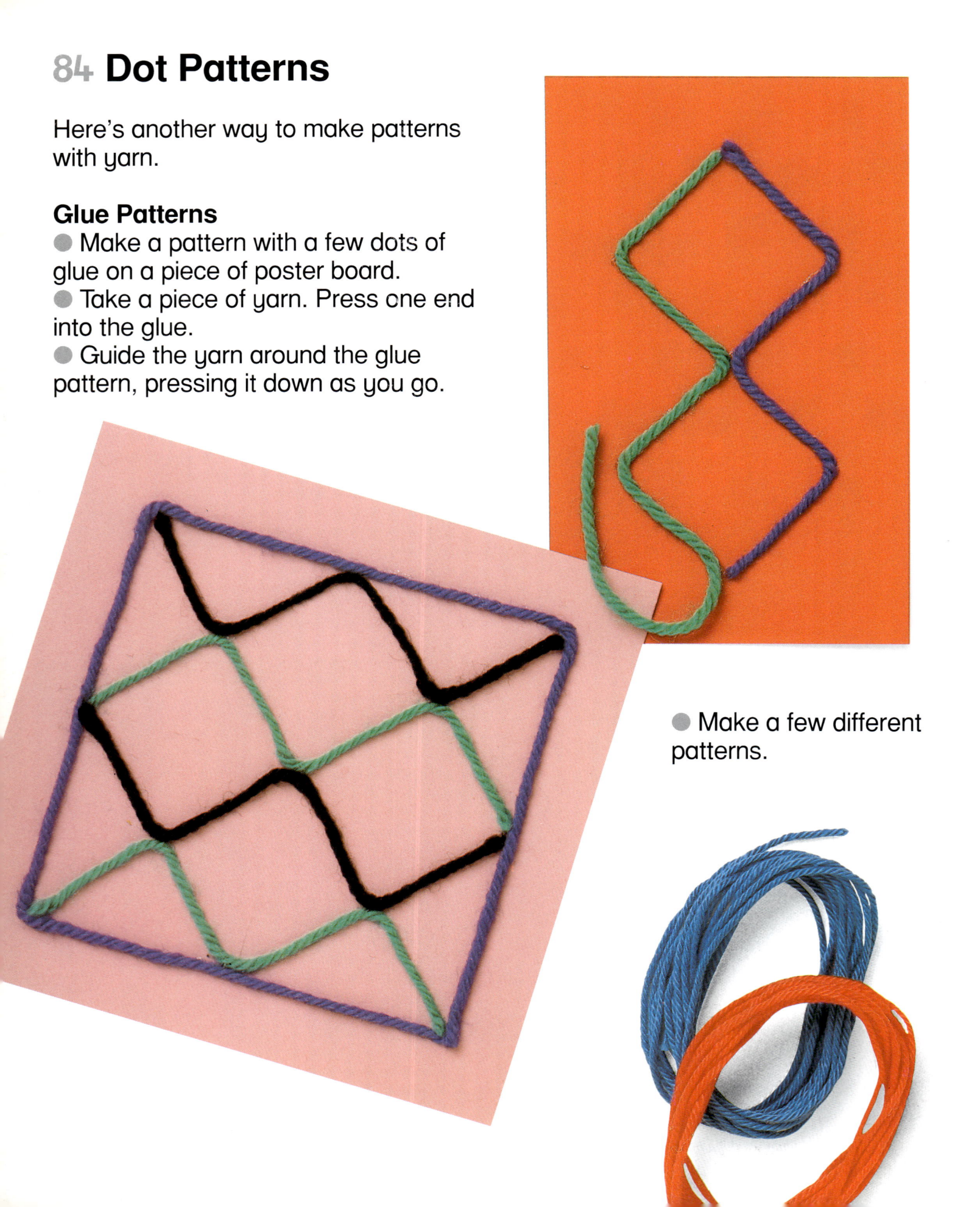

Here's another way to make patterns with yarn.

Glue Patterns
- Make a pattern with a few dots of glue on a piece of poster board.
- Take a piece of yarn. Press one end into the glue.
- Guide the yarn around the glue pattern, pressing it down as you go.
- Make a few different patterns.

Pin Patterns
- Arrange a pattern of pins on a piece of strong cardboard.
- Take a piece of yarn and tie it carefully to one of the outside pins.  
- Stretch the yarn around the pins, twisting it to keep it in place. When you reach the edge, tie the end of the yarn and cut off any extra.

Here's what you learn:
- how to create dot patterns.
- how to create shapes.

86 Cut Paper

You can make some amazing patterns by folding and cutting paper.

Fold and Cut
● Fold a piece of paper in half and then in half again. Cut a small piece out of one edge. Unfold the paper.

● To make a more complicated pattern, make several cuts along the edges before unfolding the paper.

Accordion Folds
● Cut a long strip of paper. Fold it backward and forward, so the strip is like an accordion.
● Ask an adult to help cut shapes out of the folded paper.
● Unfold it to see your pattern.

Here's what you learn:
● how to create patterns.
● how to discover symmetrical patterns.

88 Tiles

Each of these tiles has a very simple design, but you can arrange them to make all kinds of patterns.

Designing the Tiles
● Cut out some squares of cardboard, all the same size.
● Choose a simple design and paint all the squares exactly the same.

Arranging the Tiles

● Start by arranging four tiles, then add some more. How many different patterns can you make with your tiles?
● Make three rows with all the tiles facing the same way.
● Make one row of tiles facing one way, then the next row facing the other way.

Here's what you learn:
● how to create shapes.
● how to create patterns.

90 Mosaics

These colored shapes fit together to make patterns.

Making the Shapes
Cut some shapes from colored paper or poster board. You could copy the shapes shown on this page.

Fitting Together
- Sort out the shapes. Put all the triangles together, all the diamonds, and so on.
- See how the shapes that are the same fit together. Use different colors to make a pattern.
- Now try fitting two different shapes together. Which shapes fit well?

Here's what you learn:
- how to create patterns.
- how to fit shapes together.

92 Wrapping Paper

Make wrapping paper by decorating paper with a pattern. Pick a shape and repeat it lots of times.

String Blocks
- Glue a long piece of thick string onto a piece of cardboard and let it dry.
- Dip the string into paint and press it onto a sheet of paper.

Potato Prints
- Think of a simple shape and draw it on a piece of paper. Ask an adult to cut your shape from half of a potato so that the shape sticks up.
- Use a paintbrush to cover the shape with thick paint. Press the potato onto a sheet of paper. Then lift it off.
- Put some more paint on the potato. Print the shape lots of times.

Stencils

- Ask an adult to help cut a stencil out of poster board.
- Place the stencil on paper. Dab paint over the stencil. Remove it carefully and repeat.

Here's what you learn:
- how to create repeating patterns.
- how to invent new designs.

94 Mirror Prints

These amazing prints are reflections.

Fold and Paint
- Fold a piece of paper in half. Open it up and put a blob of paint on it.
- Fold the paper in half again and press down firmly. Then open it up.
- Make a print using different colors. Allow one color to dry before adding the next.

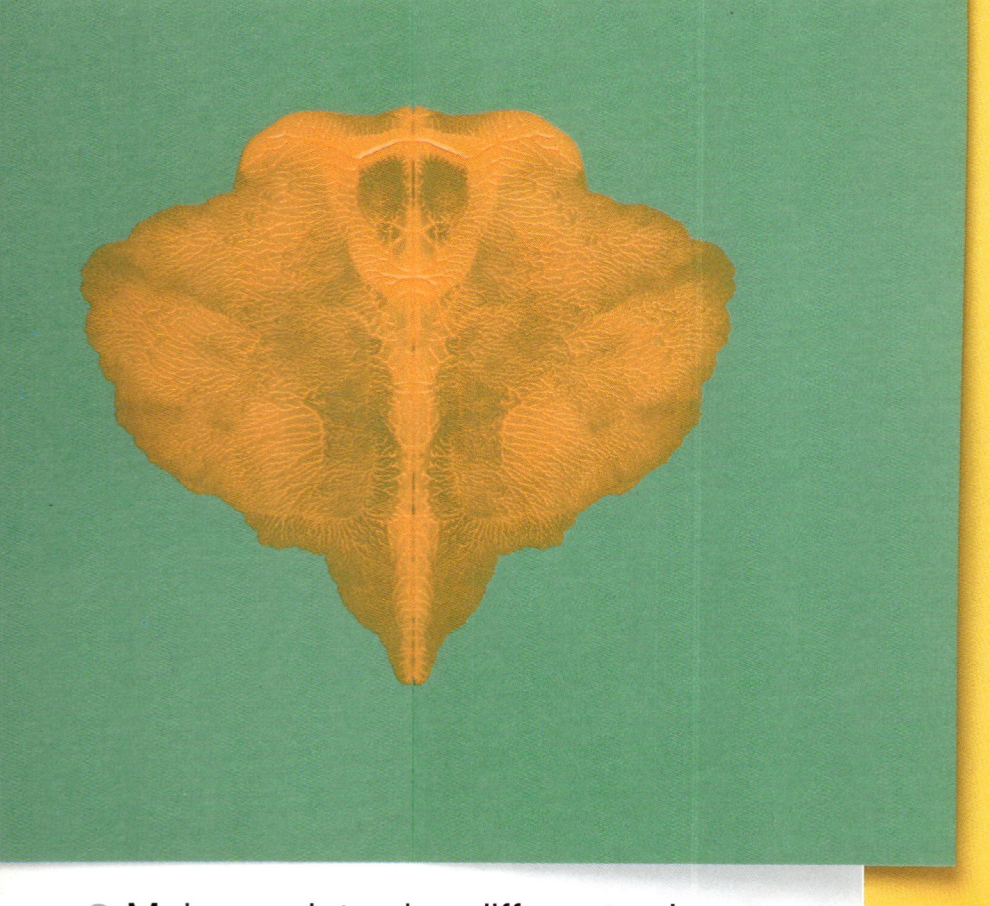

Here's what you learn:
- how to use reflective symmetry.

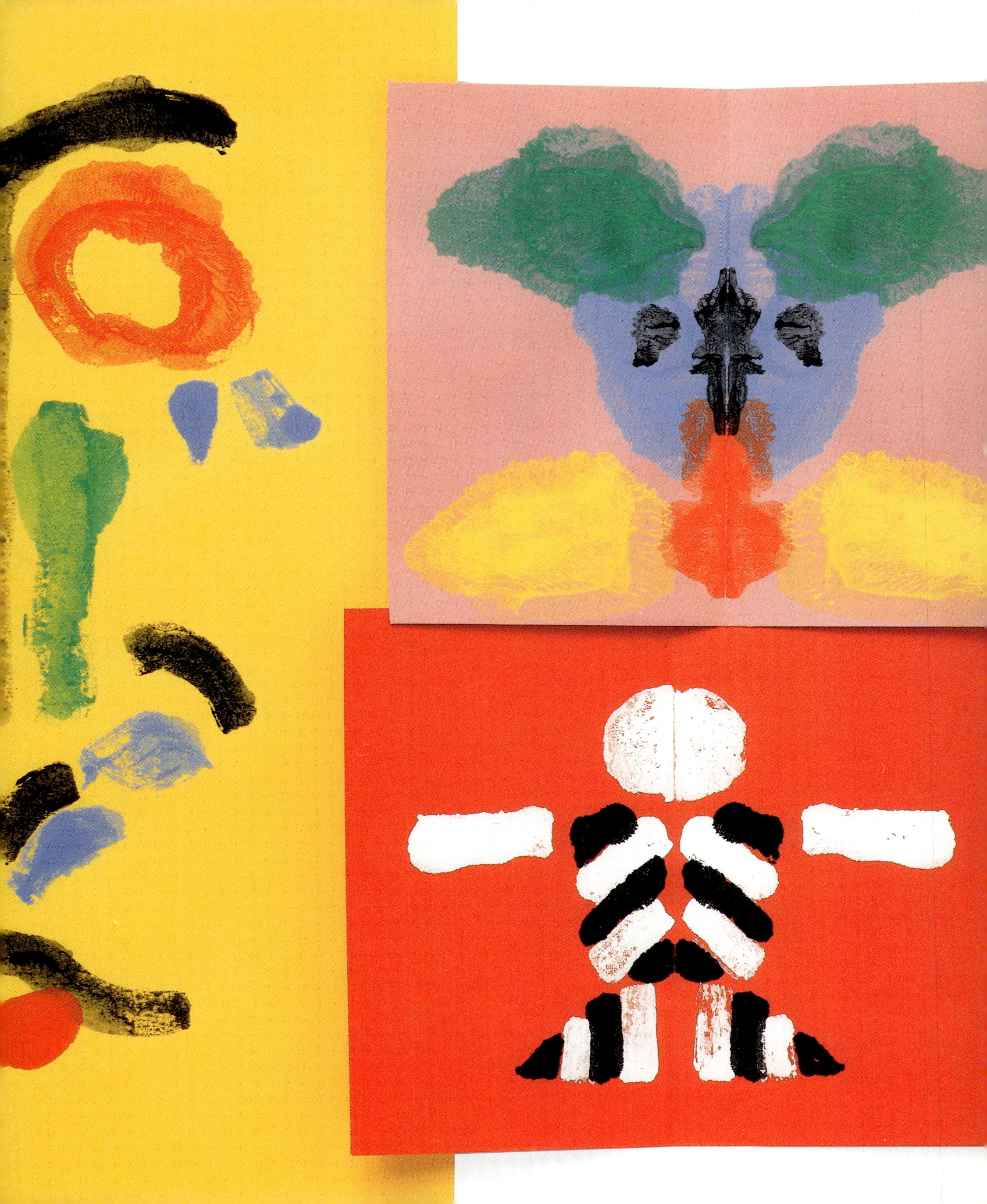

96 **Fun Clothes**

Think up some fun-looking patterns to use to decorate your T-shirts or socks. Then get permission to decorate some items. Use fabric paints, and make sure you read the instructions before you start.

Potato Patterns
- Ask an adult to cut a simple shape from half a potato.
- Cover the shape with paint, then press it onto your sock or T-shirt. Repeat to make a pattern.

Here's what you learn:
- how to create repeating patterns.
- how to use familiar shapes.
- how to invent new designs.

98 Stationery

Try using different patterns to decorate cards, writing paper, and envelopes. Use brightly colored paper and make sure you leave enough room to write! You could use some of the patterns you have found in this book or make up some new ones.

- Cut or tear shapes from colored paper. Glue them down in a pattern.
- Use plastic shapes to print.
- Cut out a cardboard stencil. Hold it down firmly and dab on paint with a sponge.

Here's what you learn:
- how to create patterns.
- how to create shapes.
- how to invent new designs.

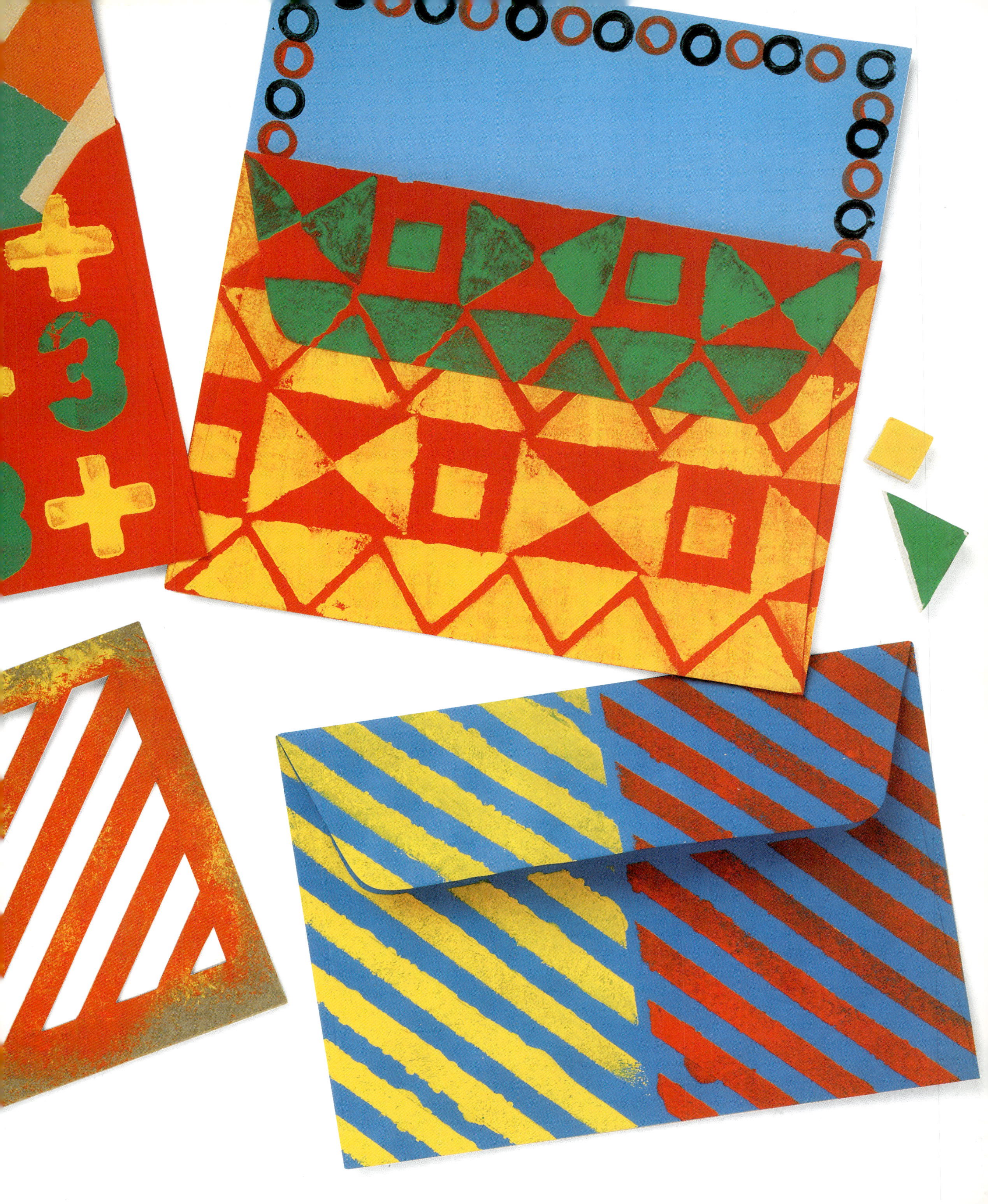

Games

102 Playing Cards

You can play lots of games with a deck of cards. Here's a way to make your own deck.

Make It – Shape Cards
- Choose four different colors of cardboard. Cut out six rectangles from each.
- Start with one set of six matching cards. Put paint on an eraser and print once on one card, twice on the next, and so on until you reach six.
- Print one to six shapes on the other sets of cards. Let them dry. When all the cards are printed, you will have a deck.

Here's what you learn:
- how to make number patterns.
- how to recognize number patterns.
- how to match numbers.

Play It – Snap!
- Deal the cards to yourself and one or two friends.
- Take turns laying a card faceup.
- When a player's card has the same number as the card that is on top of the pile, that player shouts SNAP! The player wins all the cards on the table.
- If one player runs out of cards, the others keep laying down their cards.
- The game ends when one player wins all the cards.

Make It – Number Cards
- Make a deck of cards with numbers like these. Choose four colors of cardboard and cut nine rectangles out of each color. Paint the numbers 1 to 9 on each set of cards.

Play It – Simple Rummy
You can play this game with two to four players.
- Deal four cards to each player.
- Put the rest of the deck facedown on the table. Turn the top card over and put it next to the deck.
- The aim is to collect a set of three cards. The set could have numbers next to one another – like 1, 2, 3, or 4, 5, 6 – or the set could have three of the same number.

- The first player picks up a card from the deck or takes the card that is faceup.
- The player then throws away one of the cards from his or her hand by putting it faceup on the pile.
- When one player has a set of three cards, he or she lays them on the table and shouts RUMMY!

Here's what you learn:
- how to write numbers.
- how to recognize numbers.
- how to match numbers.
- how to put numbers in order.

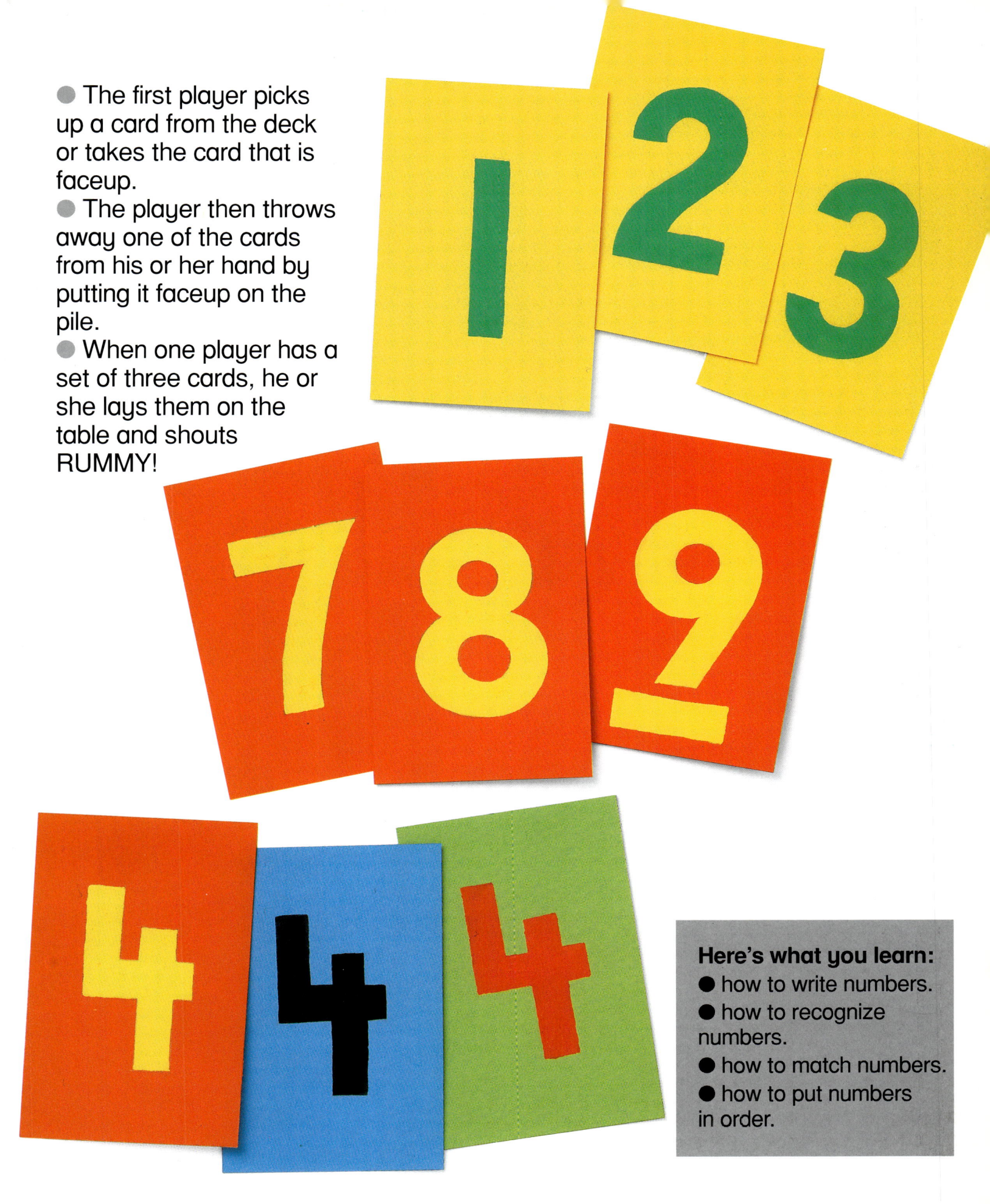

106 Memory Game

It's easy to make this memory game and fun to play it. Any number of people can play together.

Make It – Concentration

● You will need nine paper cups. Each cup should have a completely different pattern on it. You could use paint or colored paper to decorate them.
● When the paint is dry, send everyone out of the room while someone who is not playing prepares the game.
● Place two candies under two of the cups, three under another two, four under two more, and five under two more. You can put any number you like under the last cup.

Play It – Concentration

● When the game is ready, everyone comes back into the room. Take turns lifting up two cups.

● If the number of candies matches, take the candies and remove the cups from the game. Don't eat the candies yet!

● If the number of candies does not match, put the cups back.

● The winner is the player with the most candies at the end.

Here's what you learn:
● how to recognize number patterns.
● how to match numbers.

108 Bingo

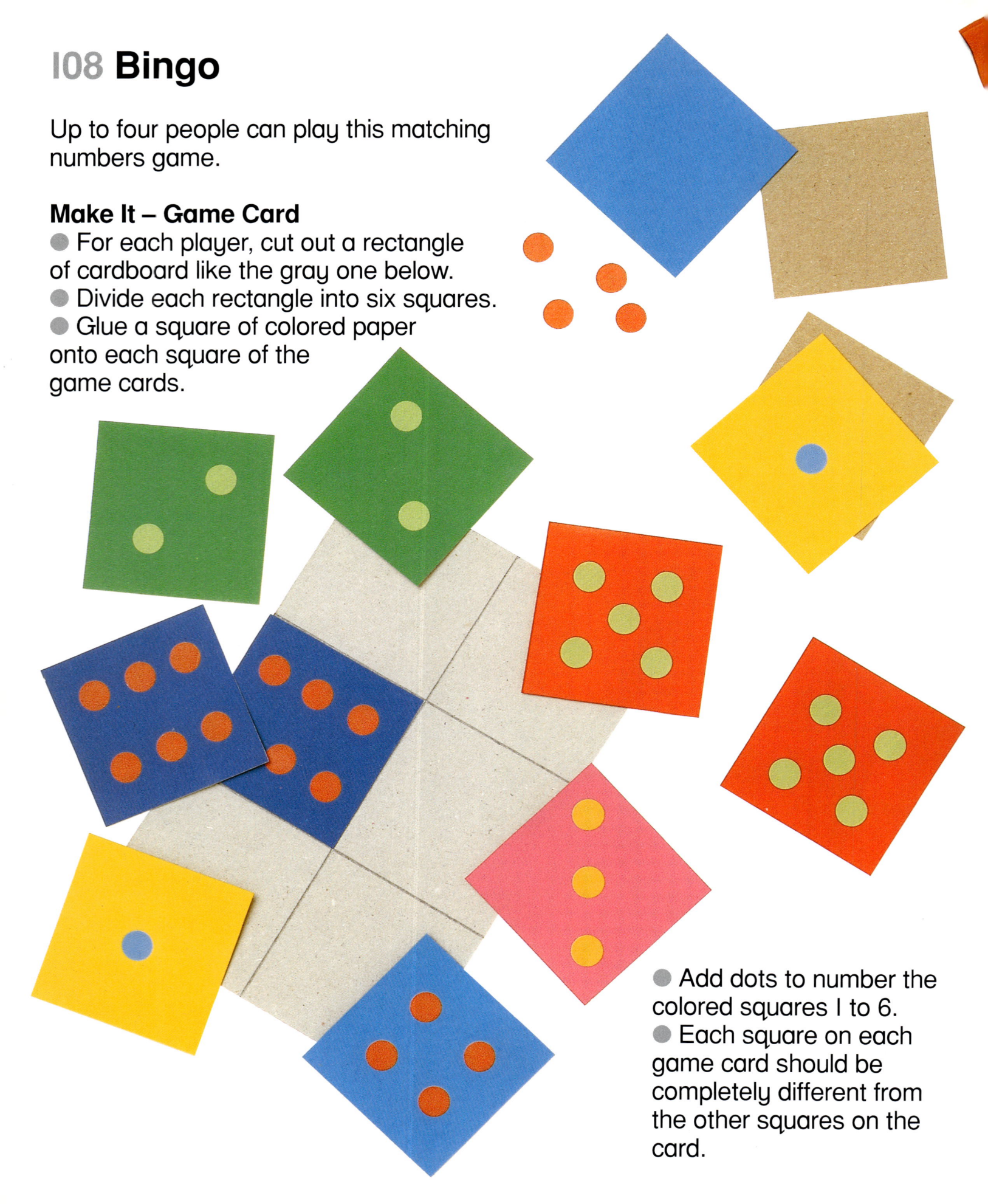

Up to four people can play this matching numbers game.

Make It – Game Card
- For each player, cut out a rectangle of cardboard like the gray one below.
- Divide each rectangle into six squares.
- Glue a square of colored paper onto each square of the game cards.
- Add dots to number the colored squares 1 to 6.
- Each square on each game card should be completely different from the other squares on the card.

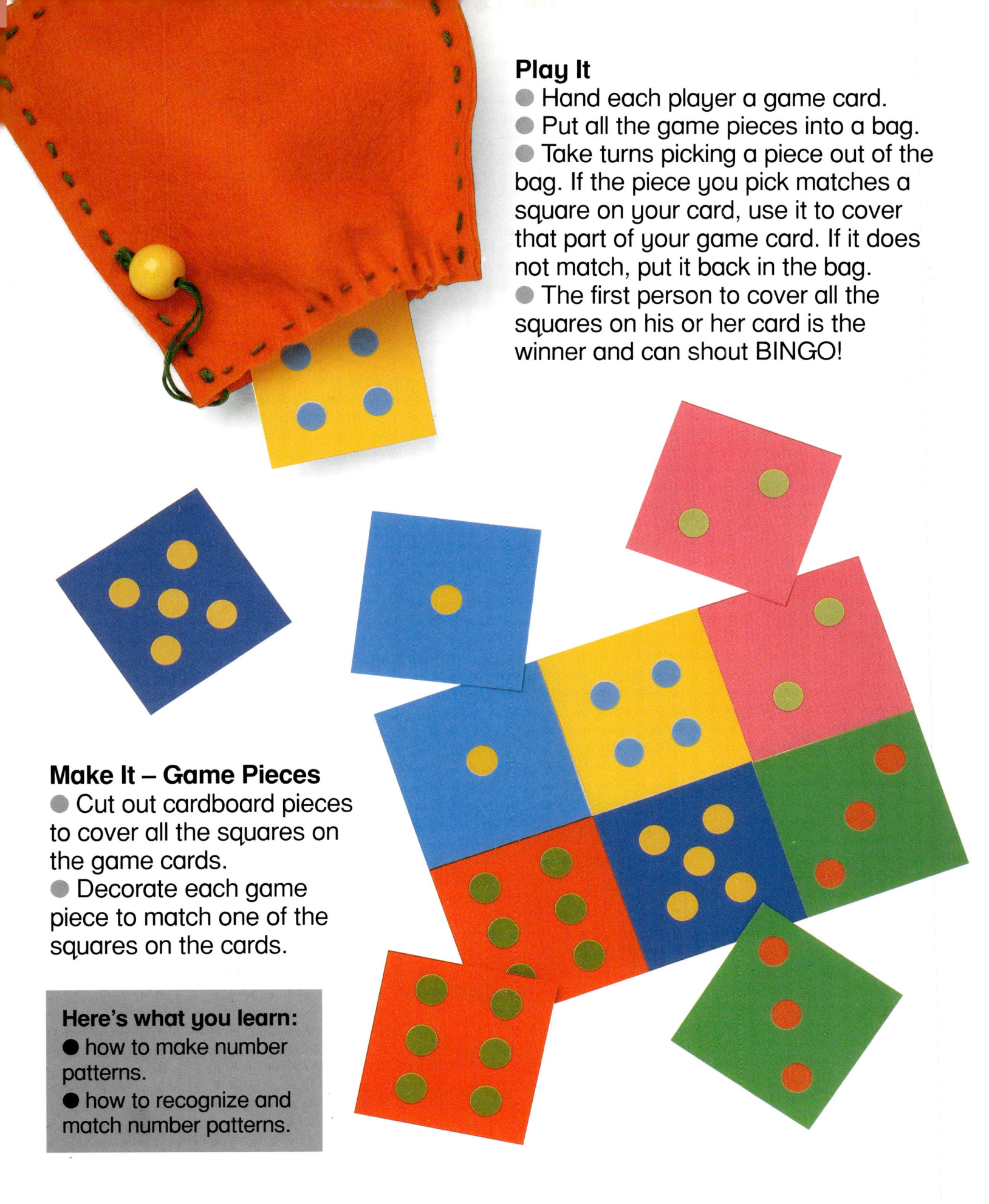

Play It

● Hand each player a game card.
● Put all the game pieces into a bag.
● Take turns picking a piece out of the bag. If the piece you pick matches a square on your card, use it to cover that part of your game card. If it does not match, put it back in the bag.
● The first person to cover all the squares on his or her card is the winner and can shout BINGO!

Make It – Game Pieces

● Cut out cardboard pieces to cover all the squares on the game cards.
● Decorate each game piece to match one of the squares on the cards.

Here's what you learn:
● how to make number patterns.
● how to recognize and match number patterns.

Make It – Pattern Bingo

● Make the game cards and pieces in the same way as before, but this time use patterns instead of numbers. We made our patterns from colored paper, but you could paint or print them.

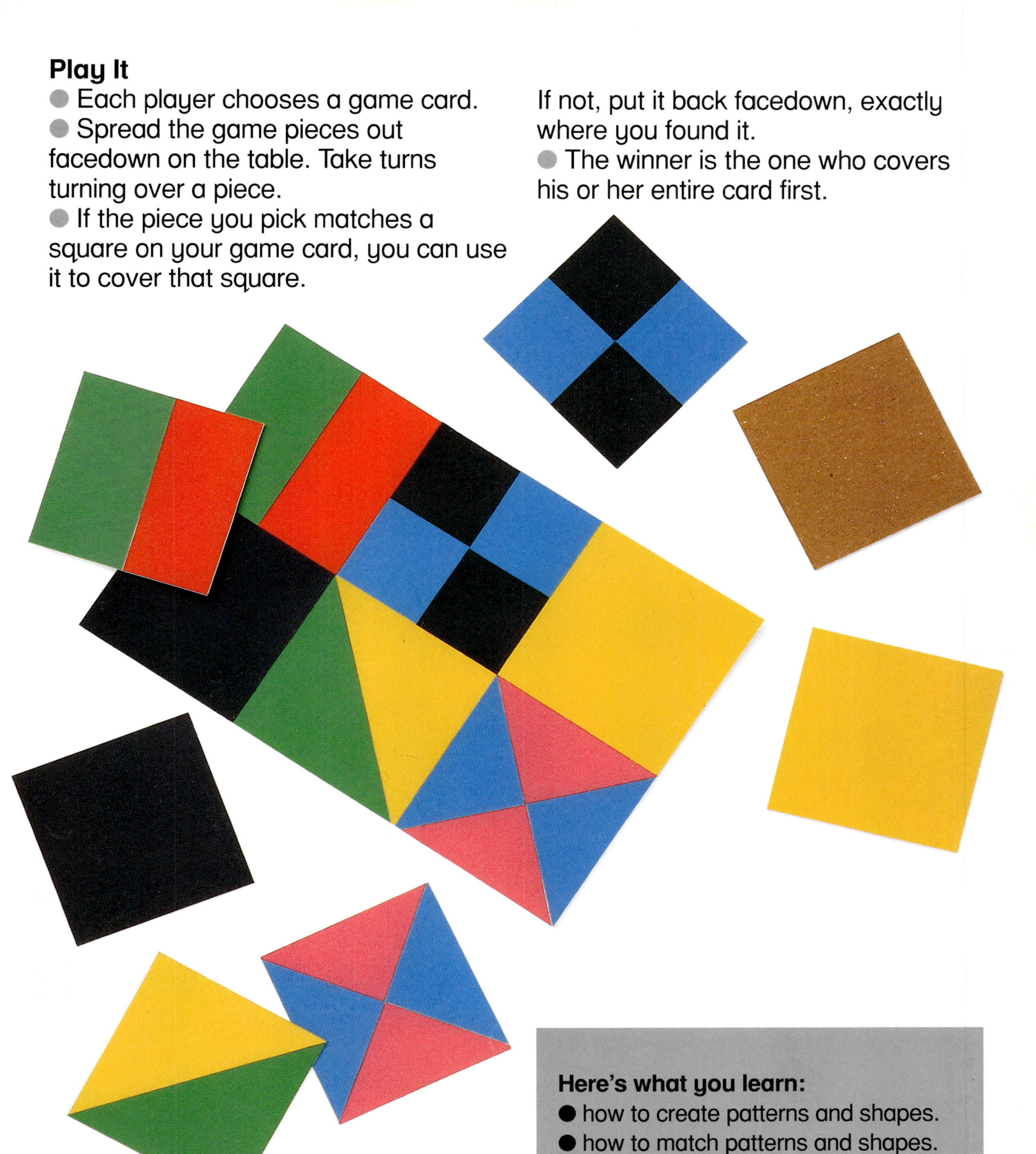

Play It

- Each player chooses a game card.
- Spread the game pieces out facedown on the table. Take turns turning over a piece.
- If the piece you pick matches a square on your game card, you can use it to cover that square. If not, put it back facedown, exactly where you found it.
- The winner is the one who covers his or her entire card first.

Here's what you learn:
- how to create patterns and shapes.
- how to match patterns and shapes.

112 Dominoes

Make a set of dominoes. The ones on the opposite page are a complete set.

Make It – Number Dominoes
- Cut out twenty-eight rectangles from a piece of cardboard.
- Glue on circles of paper or paint dots to number your dominoes.
- Make sure you follow the number patterns shown opposite.

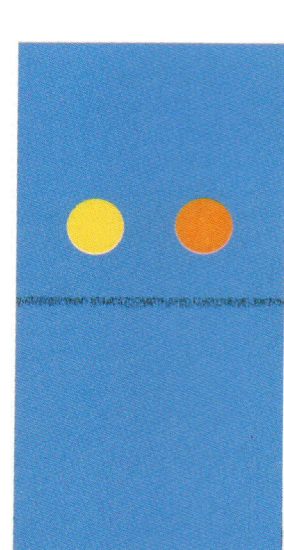

Make It – Funny Faces
You could make a set of dominoes with funny faces on them instead of numbers.
- One eye = 1
- Two eyes = 2
- Two eyes and one mouth = 3
- Two eyes, one mouth, and one nose = 4
- Two eyes, one mouth, one nose, and one eyebrow = 5
- The whole face = 6

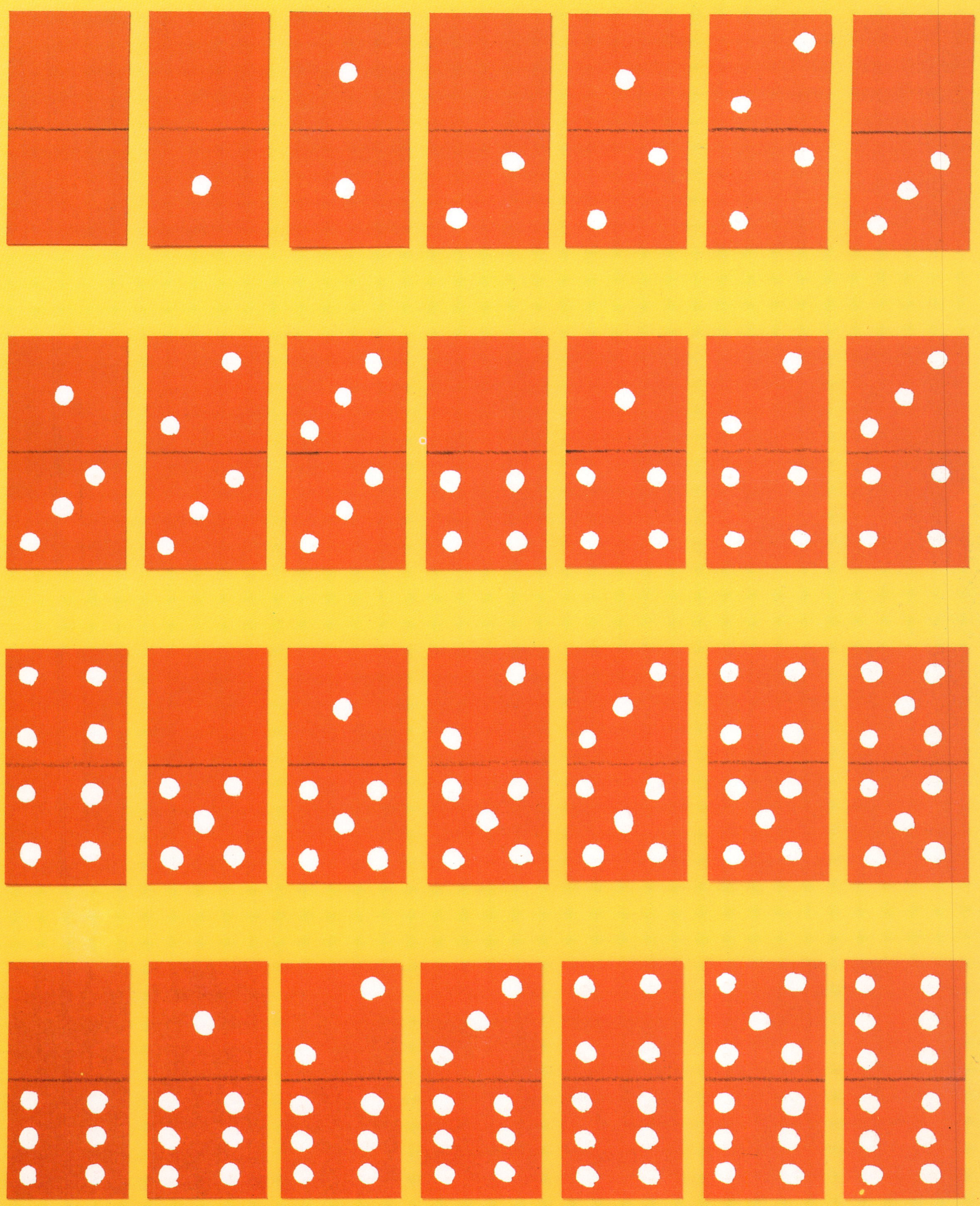

Play It

● Spread out a set of dominoes facedown on the table.

● Each player takes an equal number of dominoes: six for two players, five for three players, four for four players. Push the other dominoes to the edge of the table.

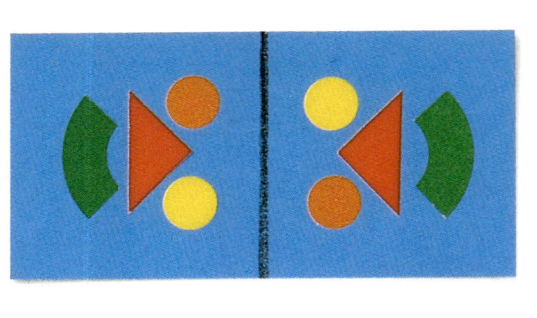

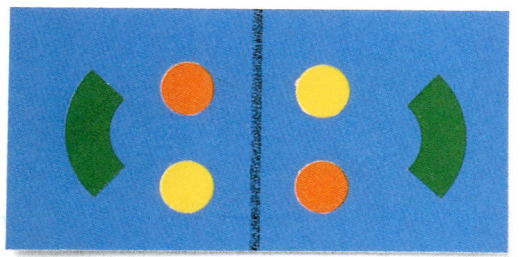

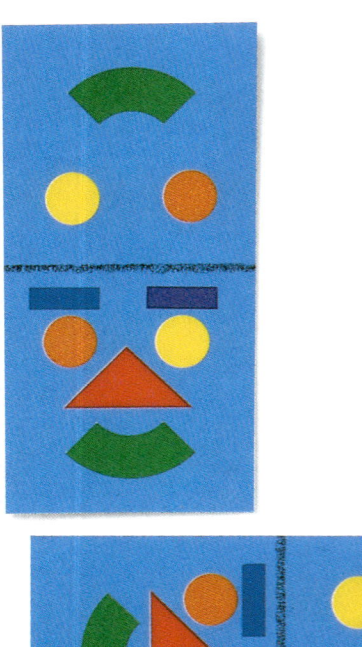

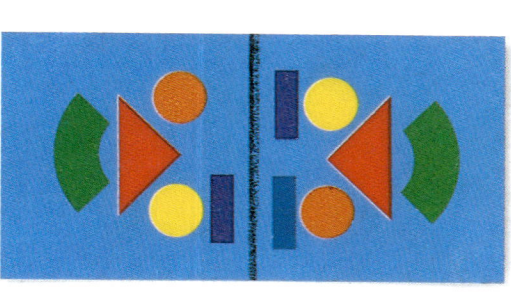

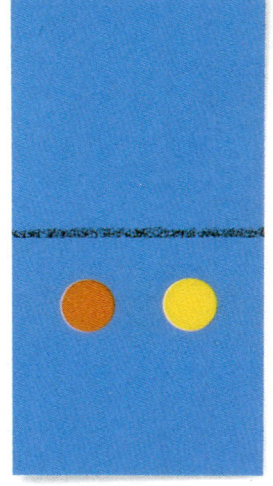

● Take turns starting the game.

● The first player places one domino faceup on the table. The next player must match one end of one of his or her dominoes with one end of the first domino.

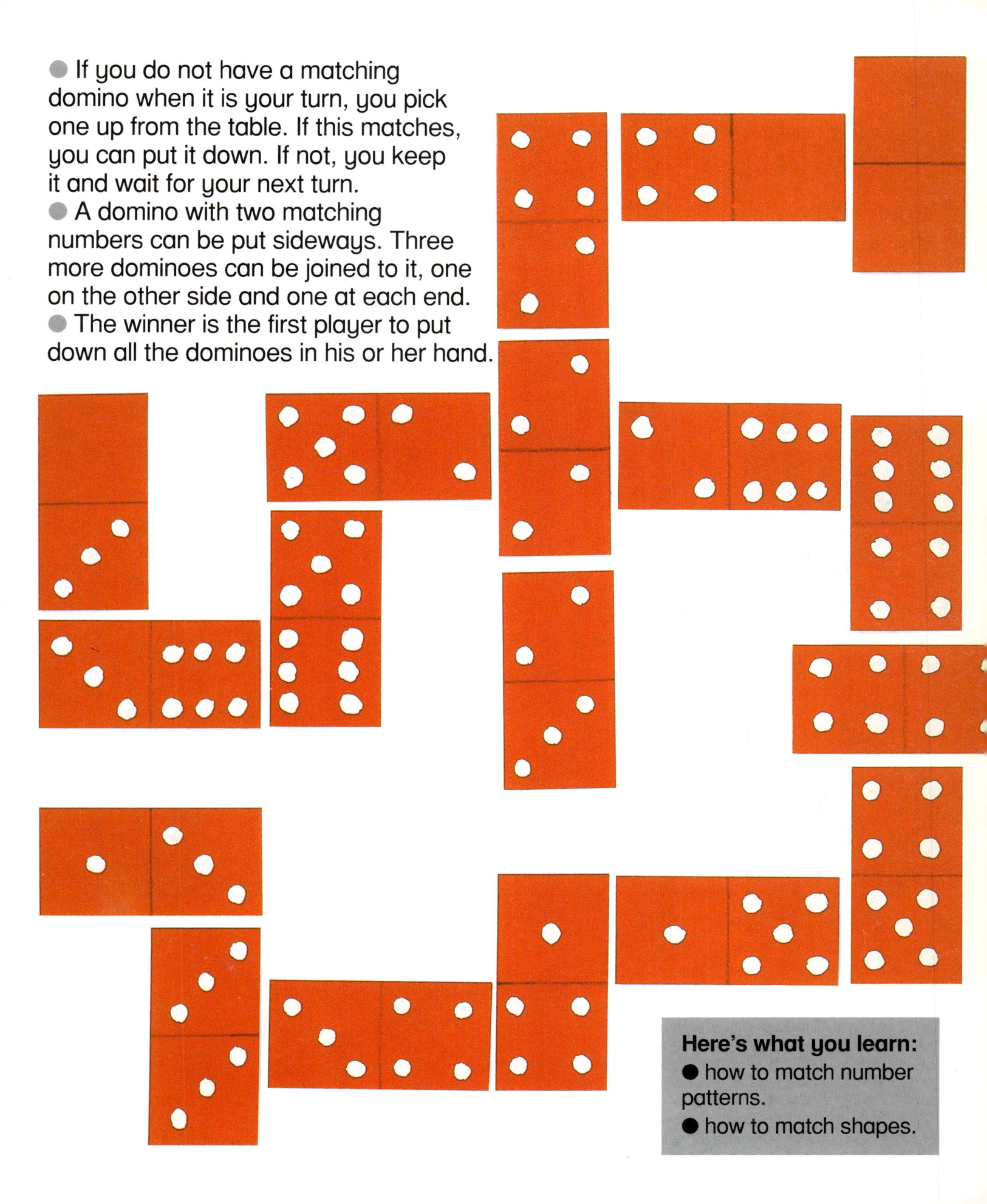

- If you do not have a matching domino when it is your turn, you pick one up from the table. If this matches, you can put it down. If not, you keep it and wait for your next turn.
- A domino with two matching numbers can be put sideways. Three more dominoes can be joined to it, one on the other side and one at each end.
- The winner is the first player to put down all the dominoes in his or her hand.

Here's what you learn:
- how to match number patterns.
- how to match shapes.

116 Spinners and Dice

For some games, you need a die or a spinner to help you pick numbers.

Make It – Dice
You can make dice out of lots of different things. Make one with the sides numbered 1 to 6.

- Find a brightly colored building block. Paint on the dots using a different color.

- Ask an adult to cut two corners off a cardboard box, such as a cereal box. Slide the pieces together and glue them to make a cube. Paint the cube and glue on paper dots.

- Make a round ball from self-hardening clay. Flatten the sides by gently pressing a ruler onto the clay. Make tiny balls from clay of another color. Press them onto the cube.

Make It – Spinners

Try using a spinner instead of a die.
- Trace around the six-sided shape on the right. Cut your tracing out, then use it to cut the same shape out of the cardboard.
- Draw lines with a pencil and ruler to divide the shape into six triangles.
- Cut out triangles of colored paper and glue them onto the spinner.
- Write 1 to 6 on the sections.
- Use a toothpick to make a hole through the spinner. Attach the toothpick to the bottom of the spinner with clay.
- Spin the shape on its stick. Which side does it rest on when it stops? That is your number.

Here's what you learn:
- how to recognize and write numbers.
- how to recognize and write number patterns.

118 Beetle Game

The object of this game is to be the first player to put together a complete beetle. Each player must have all the parts needed to make up a beetle, and you will need one die. The instructions for making the beetle are on the next page.

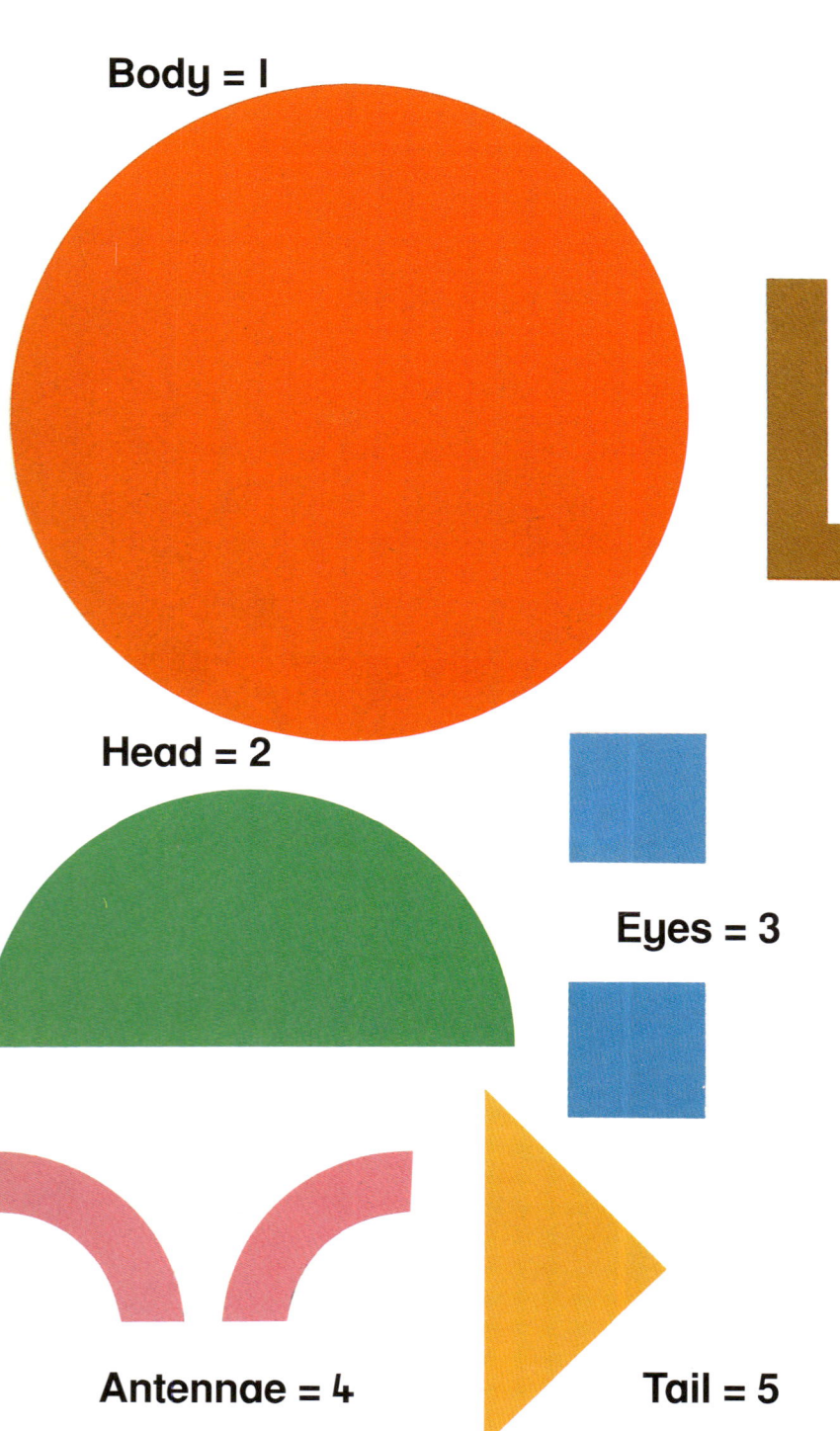

Body = 1

Head = 2

Eyes = 3

Antennae = 4

Tail = 5

Legs = 6

Play It
- Take turns throwing the die. To start, you must throw a 1 to get the beetle's body.
- Collect the other parts of the beetle when you throw the right number.
- You may not take the eyes and antennae until you have the head.
- The first player to have a complete beetle wins the game!

Here's what you learn:
- how to recognize numbers.
- about the effects of chance.

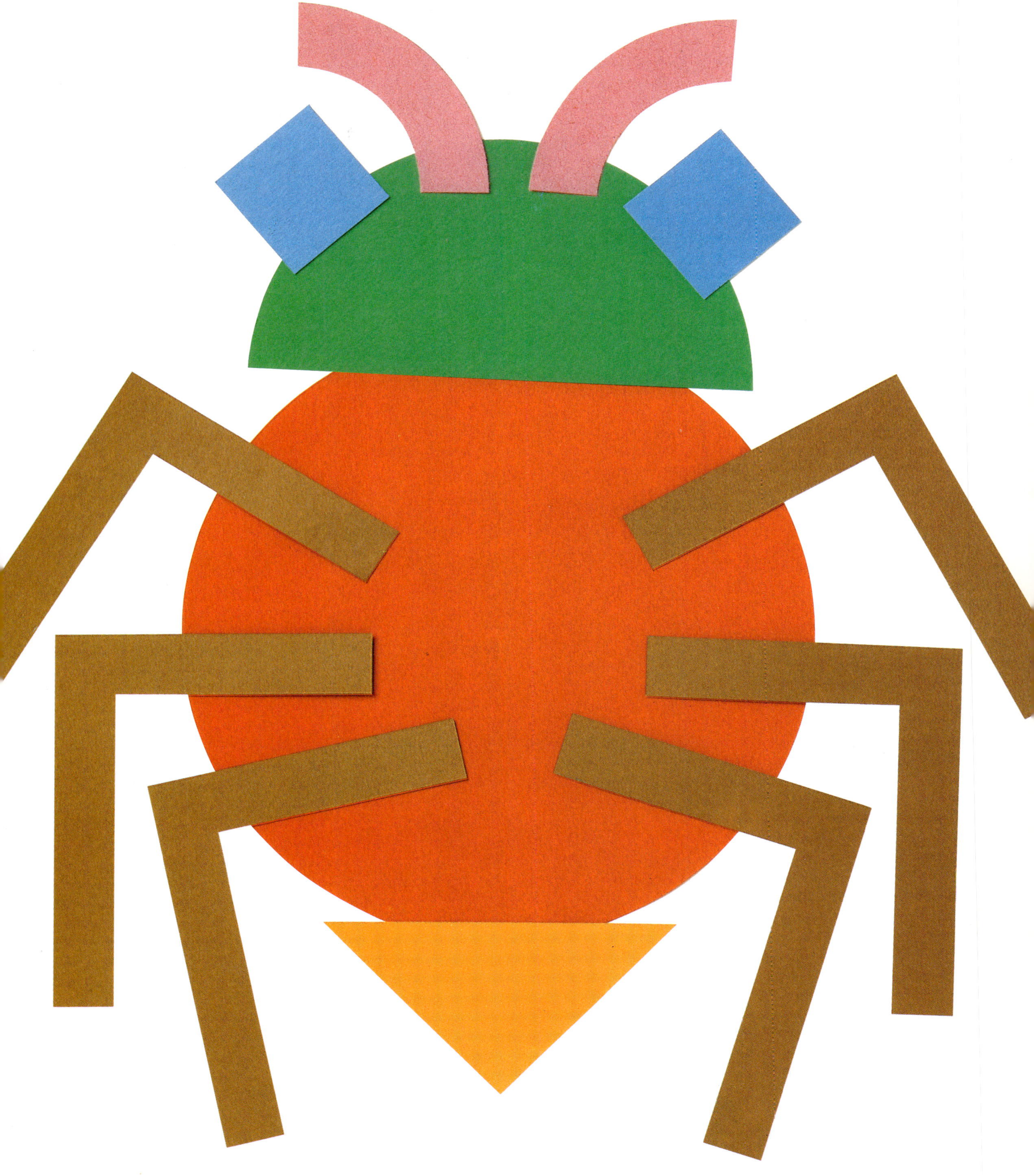

Make It – Beetle

We made our beetle from pieces of colored cardboard. All the parts can be made from squares and circles.
- You might find it easier to make your beetle by tracing around the shapes on the previous page.
- Each player needs all the parts of the beetle to play the game.

Body
- Cut a large circle from colored cardboard. You could trace around a plate.

Tail
- Cut a small square from cardboard. Make a diagonal fold so that you have two triangles. Cut along the fold. You now have two tail pieces.

Legs
- Cut a square from colored cardboard. Mark two "L" shapes inside the square as shown above. Cut out the "L" shapes to make two legs. Do the same thing with two more squares to make six legs.

Eyes
- Cut four square eyes out of the leftover squares from the legs.

Here's what you learn:
- the names of shapes.
- how to draw shapes.
- how shapes fit together.

Head
● Cut a circle smaller than the one used for the body. For example, if you traced a dinner plate for the body, you could trace a saucer for the head. Fold the circle in half and cut across the fold to make two semicircles.

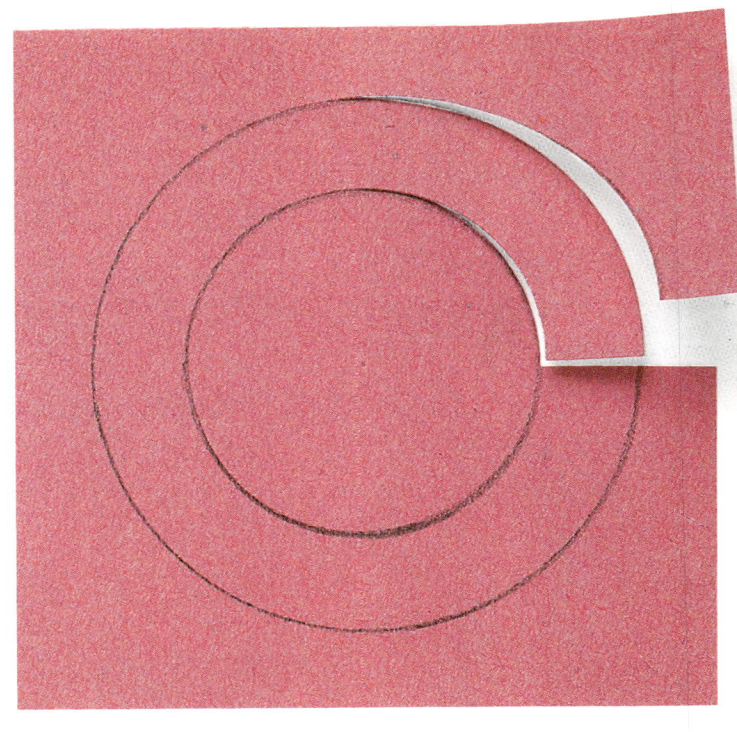

Antennae
● Use a small round object, such as a jelly jar lid, to draw around. Trace a smaller round object inside the circle you just made. Cut through the two circles as shown in the picture. Then cut around the inside and the outside circles to make a ring. Fold the ring in half and cut along the fold. Fold each section in half again and cut along the folds.

122 Board Games

Once you have made a board, you can play lots of different games.

Make It – Board
- Cut out a piece of cardboard to use as a board.
- Draw lines to divide it into squares. Glue on colored paper squares.
- Put numbers on the squares.

Make It – Playing Pieces
- For some games, each player needs a few matching playing pieces. For others, you need only one playing piece each. Use candies, buttons, corks, or shells.

Here's what you learn:
- how to recognize numbers.

Make It – Slit-and-Slot Pieces
Instead of finding playing pieces, you could make your own.
- Cut out two matching shapes from stiff paper.
- Cut a slit from the bottom of one shape to the middle.
- Cut a slit in the other shape from the top to the middle.
- Slot the two shapes together to make a playing piece that stands up!

Play It – Forward and Backward

Make a spinner like the one shown on the board below. You will also need: a board that displays numbers in ascending order, as shown below; and one playing piece for each player.

● The player with the highest blue score starts at number 1. Move your playing piece following the ascending numbers on the board.

● Blue numbers make you go forward, and green ones make you go backward.

● If you move backward off the board, you have to get a blue number before you can get on again.

● The winner is the first to go beyond number 36 on the board.

Hazards

Work your way around this board following the arrows.
- If you land on a pink square, go forward four spaces.
- If you land on a yellow circle, miss a turn.
- If you land on an orange triangle, go forward three spaces.

126 Snakes and Ladders

You need a board with colored squares to play snakes and ladders.

Make It – Snakes
● Mix two colors of modeling clay into a ball. Roll the ball into a sausage. Flatten the head slightly and add two eyes. Curve your snake into a wiggly shape.

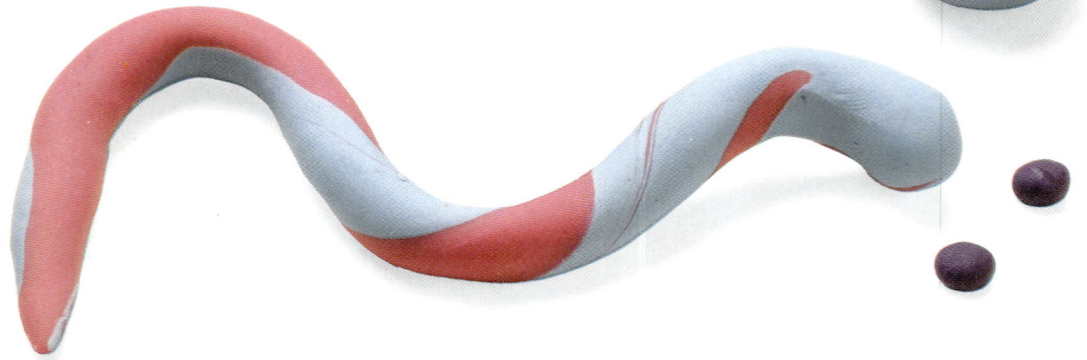

Make It – Ladders
● Cut a long strip of colored paper. Make folds along the length of the paper. Make some short and some long ladders.

Play It
● Throw a die to move. If you land at the bottom of a ladder, climb up it. If you land on a snake's head, slide down it.
● The first player to get to the end wins.

Here's what you learn:
● how to recognize numbers.
● about chance.

Index and Notes

bingo 108, 109, 110, 111
board games 122, 123, 124, 125, 126, 127
bottles 54, 55, 65
boxes 60, 61, 62, 63

card games 102, 103, 104, 105
cardboard 18, 19, 20, 28, 29, 46, 51, 52, 53, 56, 60, 61, 66, 72, 73, 80, 84, 85, 88, 90, 92, 93, 98
circles 11, 13
color 75, 78, 82, 83, 85, 94
containers 26, 28, 32
cookies 58, 59, 60, 61
cubes 26, 28, 29

dice 116, 117, 118, 124, 125, 126
doll 52, 53
dominoes 112, 113

felt 42, 43, 44, 45

glue 74, 84, 92, 98

height chart 40, 41

hexagons 14, 15

jewelry 34, 35

masks 48, 49, 50, 51
matching 104, 105, 106, 107, 108, 109, 110, 111
mobile 56, 57

numbers 104, 105, 106, 107, 108, 109, 112, 115, 116, 117, 118

oval 13

paint 9, 19, 20, 21, 30, 31, 33, 47, 50, 62, 66, 72, 74, 88, 92, 93, 94, 96, 97, 98
paper 9, 10, 11, 12, 13, 14, 15, 16, 17, 20, 34, 35, 40, 41, 46, 47, 50, 51, 52, 53, 62, 63, 64, 65, 72, 73, 74, 82, 83, 86, 87, 90, 92, 93, 94, 98
patchwork 14, 15
patterns 7, 8, 14, 15, 33
playing pieces 122, 123, 124, 125, 126, 127

printing 8, 9, 92, 93, 96, 97, 98
puppets 42, 43, 44, 45, 46, 47

rectangles 13
reflection 86, 87, 94

sand timer 64, 65
sets 102, 103, 104, 105
shapes 70, 73, 78, 90, 91, 92, 96, 98, 110, 111, 112, 120, 121
slit and slot 18, 19, 20, 21, 22, 23
sorting 75, 78, 90
spinners 116, 117, 118, 124, 125, 126
squares 10, 13
string 49, 56, 57

tangram 66, 67
triangles 10, 13, 32
tubes 21, 22, 30, 31, 32, 33

vegetables 24, 25

weaving 80, 81, 82, 83

Shapes skills index

draw and cut precisely 10-11, 28-29

fit shapes together 8, 12, 14

investigate symmetry 16-17, 20-22
investigate two-dimensional and three-dimensional shapes 18-19, 26-27
investigate three-dimensional shapes 20-22, 32-33

make one shape into another 32-33
make patterns 15
make right angles 34
make things using three-dimensional shapes 24-25
make two-dimensional shapes into three-dimensional shapes 28-29, 34
match shapes 20-22

recognize different shapes 8-9, 10-11
recognize shapes in different sizes 30-31

use different shapes in design 12

Patterns skills index

change a flat shape into a three-dimensional one 73
create patterns 76, 82, 87, 89, 91, 98
  dot patterns 85
  repeating patterns 73, 93, 97
  simple patterns 79, 80
  symmetrical patterns 85, 87
create shapes 82, 85, 89, 98

fit shapes together 91

invent new designs 93, 97, 98

match similar things 74, 79

recognize patterns 72

sort things into groups 74, 79

use familiar shapes 97
use ideas about symmetry 80
  reflective symmetry 94
  symmetrical patterns 85

Games skills index

develop spatial sense 119

explore chance 118, 126

fit shapes together 120

make and classify shapes 110
make number patterns 102, 108, 112
match numbers 103, 105, 107
match number patterns 109, 115
match patterns and shapes 111, 114

order numbers 105

recognize names of shapes 119
recognize numbers 105, 117, 118, 122, 126
recognize number patterns 103, 107, 109

write numbers 104, 117

Measure skills index

create pictures using shapes 51, 66-67

estimate area 61

investigate area 42, 52-53, 66-67
investigate area and volume 60-61
investigate size 43, 46, 56-57
investigate weight and balance 57

match shapes 43
match sizes and shapes 52-53
measure time 65
measure using units 40, 54

use measurements 40, 48
use standard measures 59